TE MANU HUNA A TĀNE

TE MANU HUNA A TĀNE

Editor Jenny Gillam
Co-editor Eugene Hansen, Maniapoto
Photographs Jenny Gillam

MASSEY UNIVERSITY PRESS

I te mārie o te waenganui pō,
ka puta te tioro a te kiwi tāne
e rapu hoa ana mōna.
Ka puta te karanga a te kiwi tāne,
‘Kei roto i tēnei ngahere tētahi hoa mōku?’
Ka mārie anō te pō i tētahi wā.
Ka pakaru te mārie o te pō i te whakautu a te kiwi wāhine,
‘E rapu hoa ana ahau, me tūtaki tāua.’
Ka whakautu a te kiwi tāne, ‘āe, me tūtaki tāua.
Mākū koe e karanga, kia mōhio ai koe kei hea au.’
Ka whakautu te kiwi wāhine,
‘Āe, ka tūtaki tāua āianei.’

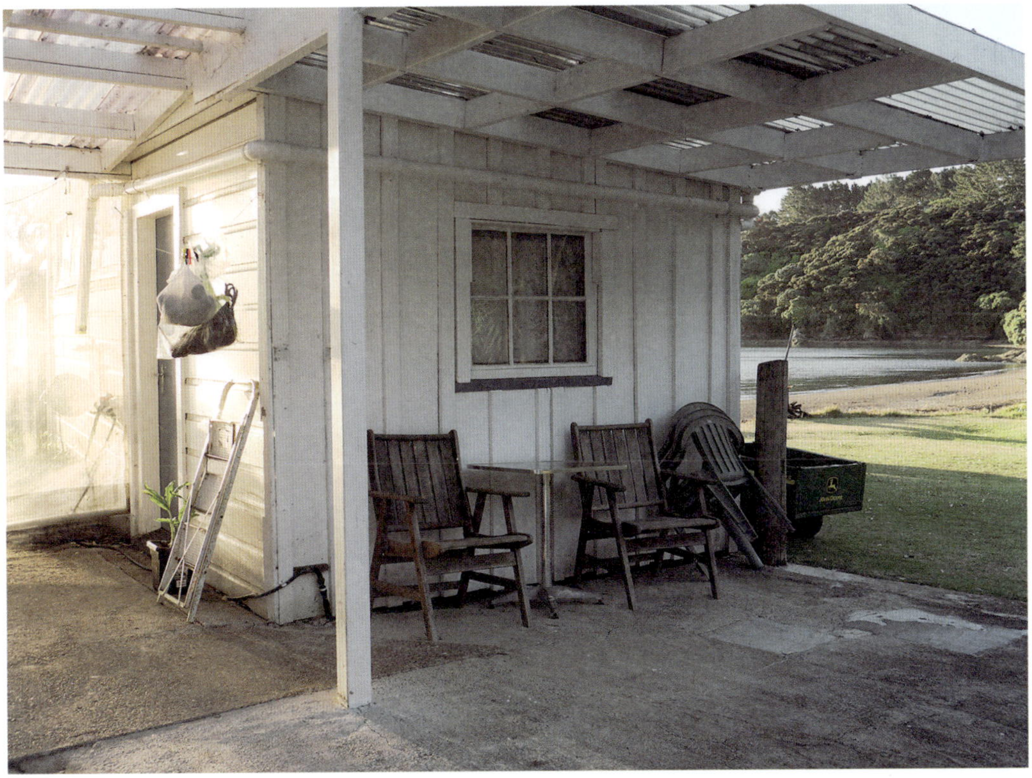

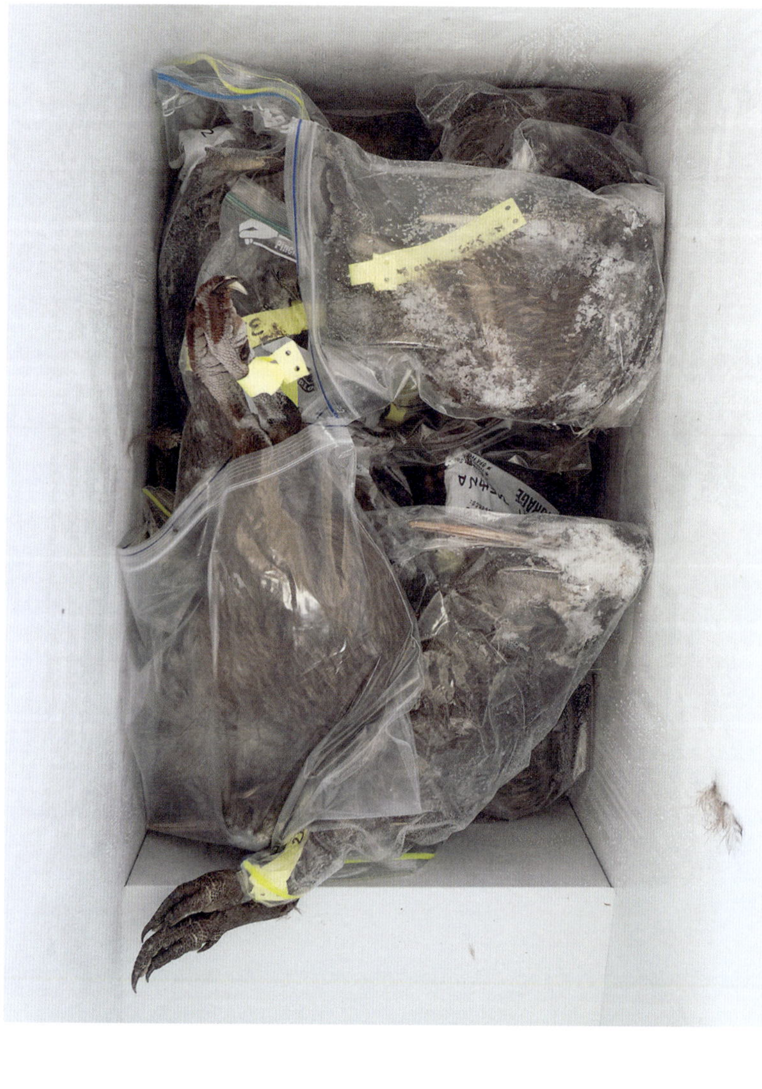

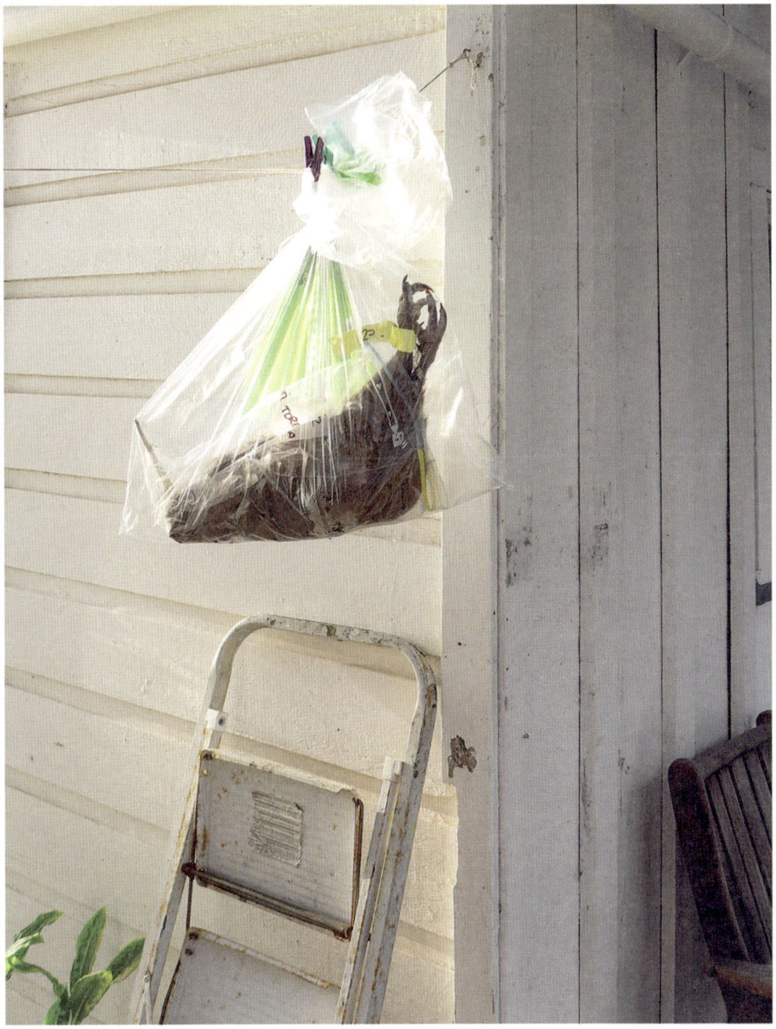

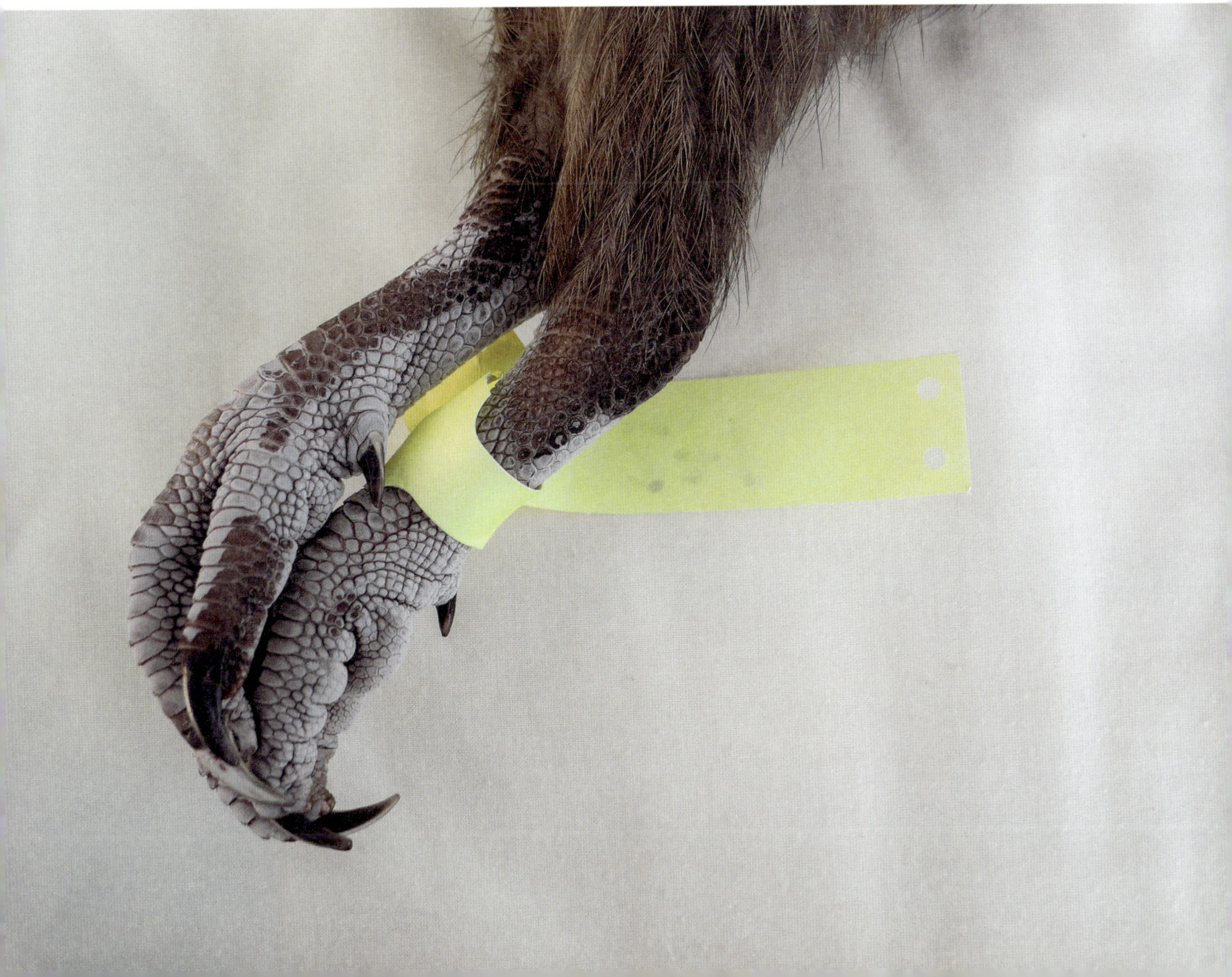

HURUHURU MANU

Raewyn Ormsby-Rihari Maniapoto, Ngāti Torehina ki Matakā by marriage
Tiwai Rihari-Rawiri Ngāti Torehina ki Matakā, Ngāti Pare, Ngāti Mau

This hīkoi for the sharing of mātauranga started with a kōrero between Hugh Rihari, kaumātua of Ngāti Torehina ki Matakā, and Rolien Elliot, manager of Te Papa Atawhai Pēwhairangi. Afterwards, Hugh told Raewyn (his wife, from Maniapoto) of the kōrero, that there were numerous native manu in freezers at Te Papa Atawhai that had been accidentally struck by vehicles or killed by dogs and handed in to them by members of the public. Native manu are protected even after death. Raewyn asked Tiwai Rihari-Rawiri, also of Ngāti Torehina ki Matakā, if she would be interested in holding wānanga huruhuru manu. Through our love of raranga we have both had experience with huruhuru manu, learning and then as assistant kaiako. With the blessing of Hugh, Raewyn approached Rolien and offered our help. Consequently, we decided on a programme of four wānanga to encompass an introduction to huruhuru manu, kiwi advocacy, pelting, and how to make muka from kōrari and attach huruhuru. For this we used non-native huruhuru.

Te Papa Atawhai record where the manu carcasses are found so that they can be returned to their rohe. We decided that two tangata from each of the rohe would be invited to participate, leaving it to them to select who would attend.

The wānanga were held on the premises of Te Papa Atawhai Pēwhairangi. Cinzia Vestena, as Biodiversity Assets Ranger, is responsible for the manu that are handed in and was an integral part of these wānanga. Her enthusiasm and participation were greatly appreciated. She and Rolien addressed Te Papa Atawhai's kiwi advocacy, including the process and requirements to apply for a permit to hold protected wildlife. This is required for all animals native to Aotearoa, and not only for living animals.

We are both permit-holders for Ngāti Torehina ki Matakā; we hold these manu on behalf of the hapū. They do not belong to us and cannot be used for commercial purposes.

Without the support of Rolien, and Cinzia's participation, the wānanga could not have happened — an example of how today we can all work well together.

Since the Waitangi Treaty settlements that have been agreed between tangata whenua and the Crown, taonga such as kiwi, kūkupa, pūkeko and many other native fauna can be returned to hapū if all legal protocols are followed. Waitangi claims are not just about whenua and pūtea, they are also about taonga and rights to our culture and customary practices. Many tangata assume the claims are driven by financial gain, but to us the process is much more; it is acknowledgement and redressing of wrongs. Before this process, government departments were autonomous. Now they interact with tangata whenua, hence the commitment to returning these manu to hapū. Te Papa Atawhai Pēwhairangi has a very good working relationship with Ngāti Torehina ki Matakā. It is a pleasure to work with tangata of this calibre.

One of the aims of these wānanga was to encourage the young to participate in something new, learning a skill that could take them into other fields of endeavour, but few took up the opportunity. It was mainly those in their thirties and older who attended, some, never having had the opportunity to learn raranga, gladly doing something they had thought would never be available to them.

Removing the pelt is not the end of the process. The pelt is then cleaned of all flesh that may have been left on it. This is less likely to be needed when pelting fresh manu. Those that have met a violent end, from dog attacks and car strikes, are more likely to need cleaning. The pelts are then hung to air-dry. This can take several days depending on the weather. Once dried they are sealed in a plastic bag with the huruhuru side folded outwards, and refrozen, as this keeps them from becoming infested by moths and other insects that would be only too pleased to have a free lunch. At a later date they can be removed from the freezer, thawed, and used for raranga. We are lucky to have freezers today, imagining the working conditions of our ancestors.

The rest of the manu remains are returned to Papatūānuku, buried with due ceremony with or without the bones. If we have a tohunga whakairo requesting bones, they are cut out and handed to them with an appropriate karakia. The bones are buried until they become defleshed. If the tohunga whakairo has access to a beach, the bones may be placed in a mesh bag and buried in the sand for sea lice to clean. They can then be used for making various items such as ngira, mau taringa and nguru. The timo can be used as autui.

The pelting — skinning — of manu is a great help to kaiwhatu. When making a kākahu, two or more huruhuru are placed one on top of the other, so keeping them in their natural position means we need not search through loose huruhuru to be able to put them together in bundles as we weave.

Most kaiwhatu are wāhine and hold the mātauranga raranga for their hapū. However, we will teach anyone who wishes to learn this skill — tāne and wāhine,

Māori and tauiwi. Tauiwi are not able to use native huruhuru in their weaving and cannot obtain a permit to hold native wildlife, but they can pelt introduced manu.

In the past, mātauranga was generally only passed to a select few, if passed on at all. For everyday purposes, skills held within a hapū would be used. But for ceremonial garments, carvings, weapons or adornment a superior talent would be sought and a trade of some kind would be arranged. For example, at the time when the National Marae at Waitangi was built, Ngāpuhi were not strong in whakairo so tohunga whakairo from outside the rohe were brought in.

Resources would also be negotiated among hapū. Inland iwi would come to the coast to gather kaimoana at certain times of the year. Special areas were set aside for this and the hapū who were granted this privilege were able to stay for limited lengths of time to gather and dry kaimoana. Outside of these sanctioned periods they would be considered a raiding party and treated as such. Reciprocal rights were given to the hosting hapū. At specified times they could hunt manu and gather kai and rongoā in the inland rohe, thus a sophisticated form of trade was carried out. Huruhuru would be traded during the course of these expeditions.

We teach wāhine so that they can take the mātauranga back to their marae and teach their wāhine, ensuring the mātauranga is not lost. Teaching wāhine is a form of empowerment. In a society where gender inequality is still the norm we feel it is important that wāhine have their own forms of mātauranga, and that they can make their own decisions about to whom they wish to impart the mātauranga. And by empowering wāhine we feel it can contribute to the empowerment of the whole hapū.

Becoming involved in this kaupapa has been a way of giving back — passing on the knowledge we feel lucky to have been taught. This is not an onerous task for Raewyn as it is something she enjoys. Tiwai does not like the pelting so much, but sees it as a necessary part of revitalising traditional raranga practices. This revitalisation and the preservation of materials has been ongoing. If we think about Toi's account later in this book of the first national hui raranga held in 1983, there was concern for the preservation of plants used in raranga. We are grateful that we continue to have access to them today, although, sadly, there are some that are still hard to come by in a quantity that is sufficient to be of use.

This set of photos is of a wānanga for the members of Ngāti Torehina ki Matakā held at the whare of Raewyn and Hugh. We pelted eight kiwi that morning. Like all such things, only a few are interested or can cope with the pelting process. Jenny, Raewyn's daughter-in-law, was visiting and was happy to document our hui for us. And from this, Ngāti Torehina ki Matakā now has a pictorial record that can be passed down.

Tiwai is the daughter of a kaiwhatu, so maybe this is in her DNA; she also carves and paints. She went to kuratini to learn more and advance her knowledge, earning a tohu mātauranga. She weaves kete of all shapes, sizes and colours, making them useful for a variety of purposes, as well as kākahu, whāriki, mau taringa and putiputi, passing the mātauranga to others by holding regular wānanga.

Raewyn's mother was Pākehā so she did not learn raranga until relatively recently, but she has a background in weaving with looms of varying sizes and shapes using wool, mohair and other textiles. Her preference is for tāniko (sometimes likened to embroidery, tapestry and braiding, but without the use of a needle). Tāniko is made from lengths of cotton, wool or muka of two or more colours to form a pattern and is used in the making of kete and the borders on kākahu. It can also be used as art pieces and wall hangings, with or without huruhuru. The variety of things that can be made by raranga and tāniko is endless, as are the materials that can be used. Contemporary raranga, with the application of mātauranga, is only limited by the imagination of the kaiwhatu.

When making a kahu kiwi we always use kōrari. The body of the kahu kiwi is made in the traditional way — no contemporary products are used. The whenu is muka, as is the aho. The tāniko that adorns the top and sometimes the sides and bottom edges of the kākahu can be made using dyed muka, although it is more usual today for it to be made of cotton. While we now have access to huruhuru kiwi, the process is lengthy and it takes some time to bring together enough helpers to complete a kahu kiwi.

For Raewyn, there are mixed feelings in handling these manu that come to us, including immense sadness at the death of all these manu that are today classed as taonga. Can you imagine the surprise of our ancestors at this whakaaro? To them, kiwi were hunted for kai and the byproducts used to produce a range of things — the huruhuru for kākahu and whakapaipai; the wheua for matau, kōauau pongāihu, pere manu and many other items. Can you imagine their amazement at the situation that kiwi are now in as we fight very hard with pest eradication, predator and dog control, trying to protect their habitat? Living as we do in rural Pēwhairangi this is ongoing. In this part of Aotearoa we live with the largest population of North Island brown kiwi. We see kiwi wandering all over the place at night. If we are out and about, we see or hear them frequently. They do not fear us, do not recognise us as a danger, and will walk right up to us as long as we remain still. Balancing the sadness we feel, their death enables us to make a kahu kiwi that will enhance the mana of those who are privileged enough to wear it. Not everyone is given this right. So in this we see the manu as living again in ātaahua.

TE KAITIAKI O NGĀ TIKANGA TOI RARANGA

Toi Te Rito Maihi Ngāi Te Ipu, Ngāi Te Apatu o Ngāti Kahungunu, Ngāpuhi

I was fascinated when, aged seven, I first saw my father weaving a kete. He would throw the kete behind him as soon as he saw me. As I was fair skinned he was determined I would not do anything that marked me as being Māori. This was despite the prediction made at my birth — 'She will bring the mana back to raranga' — by the group of kuia waiting outside for my father to show me to them. They asked what he had named me. 'Toi'. And as his surname was Te Rito the implications were clear. Apparently when they became too excited my mother grew anxious and asked Dad to bring me back.

At that time it was not unusual for Māori parents to encourage their children to take on Pākehā culture and to minimise their own kaupapa in an attempt to thrive in a colonised world. My father was clearly trying to live in both worlds. During my childhood I remember occasions when Dad had the time and the need to weave — we would put up with having to bathe standing in the copper due to the bath being full of kōrari soaking in water to soften it as part of its preparation for raranga.

Despite my father's resistance, and without his knowledge, my Pākehā mother, with her terrible accent, had taught us a little te reo Māori. She thought that if we won a scholarship to a Māori college, his pride would not let him turn it down. So, on my twelfth birthday I started at Queen Victoria School, an Anglican boarding school for Māori girls in Auckland. Here I met my lifelong friend Connie Tuaupiki. She was fluent in both te reo and English. To my joy, we were introduced to tāniko, a twining technique used by Māori to make the patterns commonly seen on kākahu used in the past. I noticed that the girls were all wearing tāniko tātua, and I realised it was the older girls who were teaching the younger girls. The patterns all consisted of stripes, diagonals or diamond shapes, and yet with 80 girls there were 80 different patterns. In those days hapū would use only the patterns that their grandmothers had taught them so I was amazed to learn that there were so many variations within a hapū. I took my first piece home to my mother. She put it around the crown of a pōtae made by Tupanapana, my grandmother's sister.

As the years passed, I continued my passion for raranga — sometimes if I saw a beautiful kete being carried by someone (although this was rare in those days)

I would follow them and strike up a conversation. One kete was so unusual that I followed it in my car for some miles. When the woman stopped I introduced myself and we discovered we both had boys in the same boarding school! She showed me the beautiful whāriki woven by her mother from which she had taken the pattern.

As my interest in raranga grew I started to recognise that houses with neglected kōrari planted in their surrounds indicated that a kaiwhatu had lived there but was now too old to care for the plants. I would visit the houses, hoping that the whānau had kept some of her work and would be happy to show me. Well cared for kōrari indicated a kaiwhatu living and using their kōrari. These homes were well worth a visit, and sometimes they knew of other kaiwhatu I could visit and learn from.

What shocked me, though, was that often the tamariki didn't know their mother could weave! The need to have many tamariki to replace those lost through the land wars with the Pākehā meant that wāhine had no time for raranga. Although their mothers wove items such as whāriki for the floor and kete to hold the fruits of the garden, they had dropped these skills while busy with their own tamariki. When their tamariki left for the cities, only then could they return to raranga. But when the tamariki were visiting they would pack it away, so the tamariki didn't know their mother could weave!

When I was working as a primary school teacher, I was awarded a month's leave with pay by the Department of Education to pursue my interest. I took the opportunity to graph the tāniko patterns of two kākahu in Auckland War Memorial Museum that had long fascinated me. I was intrigued to see that the kaiwhatu had on occasion divided an aho to make an extra stitch — necessary to make the pattern. Summoned to bring my findings to show Ngāpuhi kaumātua Taru Rankin, who had, unbeknown to me, been following my pursuit of knowledge, I spread my graphs on the carpet in front of him. He spent some time looking at what I had done, then he asked me, 'What if you stepped into the pattern?' My answer: 'I'd be part of the pattern. It would stretch to infinity all around me, above and below me — I'd be part of the pattern!' He was pleased with my answer, and gestured for food to be served immediately. Later I drove home with my mind awhirl! Now I knew why I hadn't yet met anyone to teach me raranga. I had first to understand the patterns of tāniko.

The years passed. Wherever I was tutoring raranga, from Foveaux Strait to the Far North, I would spend the money I'd earned on a cheap rental car and would visit small and large museums. I would visit and listen to somebody's whaea, nana or kaumātua, or anybody else with an interest similar to mine. I found a wealth of information from these knowledge-holders, for which I have always been grateful. I appreciated their caution and their careful assessment of me as someone whom they could trust to share their knowledge.

I would learn the significance of the aho and whenu. Aho is a word for connection, which makes perfect sense when you see the aho connect the whenu. Whenu is thought to be an abbreviation of whenua — the Earth that sustains us. Whenua is also the name for the placenta that sustains us before birth. An older name for whenua is Io — the omnipotent one. So when you wear a kākahu, the Io is against your skin. And the aho, that connection with all of the elements that you live among and learn from, and the knowledge that you then hand on, it's all there when you put the kākahu on.

I became a member of MASPAC (the Māori and South Pacific Arts Council). Ngoingoi Pēwhairangi called a meeting of wāhine she knew who were involved in raranga. I was one of them. Ngoi was an exceptional, gifted wahine in many spheres and was worried about the disappearance of raranga. We discussed the need to bring kaiwhatu together so that we could hear from them about what type of support they needed.

In October 1983 we assembled at Ngoi's Pakirikiri Marae, in Tokomaru Bay. Air New Zealand donated the fares for kaiwhatu from all over Aotearoa, from the Far North to a lone representative from Stewart Island. But many came by car, accompanied by their tāne. At three thirty in the afternoon, buses, vans and dozens of cars assembled at the entrance of the marae of Te Hono ki Rarotonga. The whakaeke of the tangata whenua began, the kaikaranga of the manuhiri rang out. Before the wharenui a group of wāhine welcomed the manuhiri with poi and waiata. The huge visiting party came on to the marae — 160 had been planned for, and more than 400 came. Among them were Te Arikinui, Dame Te Atairangikaahu, the Māori queen, and a lively contingent from the Pacific Islands. The harirū was savoured, promising a productive weekend.

There were many presentations, interspersed with kai, waiata and raranga. The overarching kōrero was that you must know your materials, their use, and their conservation; and that any age is the best age to start learning, and the very best learners are those who are interested.

Emily Schuster told about the scarcity of pīngao. She was adamant that we must protect that resource for the descendants of tangata whenua.

Hine Matikotai presented about traditional tukutuku.

I showed a magnificent kākahu huruhuru, gleaming in the sun, silky in the wind, and spoke of raranga as an art form that influences all the different facets of our lives, including the more subtle aspects of spirituality and symbolic meaning.

Rene Orchiston of Gisborne felt that kaiwhatu Māori were using inferior materials, so 18 years earlier she had started cultivating a collection of over 50 varieties of harakeke, most of it sourced from kuia. With greatly appreciated assistance from Diggeress Te Kanawa, whose deft hands had slit, spliced and stripped them, Rene presented examples of some of her collection.

Digger spoke about the work she had learned from her mother, the greatly admired Rangimarie Hetet, saying, 'I am not a teacher, but a person who shows people.' Quietly, with humility, she described how Māori have had to overcome many things, sometimes forcing a need to break with tikanga, including the tapu of raranga work. She emphasised that, while we must now share the mātauranga, respect must be maintained to protect traditional knowledge. She spoke too of the difficulties of finding the appropriate plants for dyes. Listening to Digger, a tohunga of knowledge of the past, was truly moving.

Te Aue Davis discussed the conservation and restoration of woven garments and natural fibre in textiles. She spoke about her concerns surrounding the lack of attention to Māori fibres by restoration experts in our museums due to their emphasis on Western tapestries and embroideries, and that practically nothing had been done on the garments of the indigenous people of the land. She then demonstrated how to wash and repair damaged kākahu.

The Pacific Island women spoke of their reluctance to use Māori resources without permission. They asked for guidance about their use of harakeke rather than their native pandanus, and identified their difficulties with it being an unfamiliar, harder plant, moving and shaping itself differently from pandanus. They also wanted their children to know how to prepare and handle the pandanus.

By the end of the hui we had agreed that an organisation should be established to coordinate and care for the interests of Māori and Pacific Island weavers, and that the hui should strongly petition for the conservation of pīngao, kiekie, and harakeke reserves by:

- placing a rāhui on threatened areas;
- allocating Crown lands for harakeke planting;
- managing the harvesting of reserves under the supervision of local Māori.

This submission would be made to the Department of Maori Affairs, local authorities, the Commission for the Environment, the Ministry of Works and Development, and conservation organisations.

The hui also acknowledged the need of the Pacific Island weavers to have a continuing supply of pandanus.

So began the annual hui of Aotearoa Moananui-a-Kiwa Weavers, held on Labour Day weekend each year, which continues today as Te Roopu Whatu Raranga o Aotearoa.

Over time, with the hui becoming more focused on sharing techniques and understanding tikanga, the older kaiwhatu who were doing much of the teaching were becoming exhausted. In addition, they were often teaching someone who lived far from them, so if the pupil became lost when weaving they had no one to correct them and had to wait until the next annual hui for assistance. So it was agreed to have smaller regional hui.

As the northern representative, I chose Pukepoto as the site for our hui. It was ideally situated and had experienced kaiwhatu in Maata Te Maru and Tottie Robson. Other established kaiwhatu, Florrie Berghan and Lydia Smith, lived nearby in Ahipara. As Maata had taught the local girls to make kete, pōtae and whāriki, there were quite a few capable kaiwhatu.

We decided a kaiwhatu with a car from each hapū should be approached. She would choose who would accompany her and be given petrol money. Kōrari would be no problem, for there was Viv Gregory's plantation, which he had developed over the years, just across the road. For kai, we would pay a local whanaunga for a beast, and another for a couple of sheep. Another was a fisherman. He would provide kaimoana.

So it was settled. All was arranged, everyone happy.

We began alternating regional hui with national hui. We could run three regional hui for the same cost as a national hui, and we now had a great number of small groups producing very fine work.

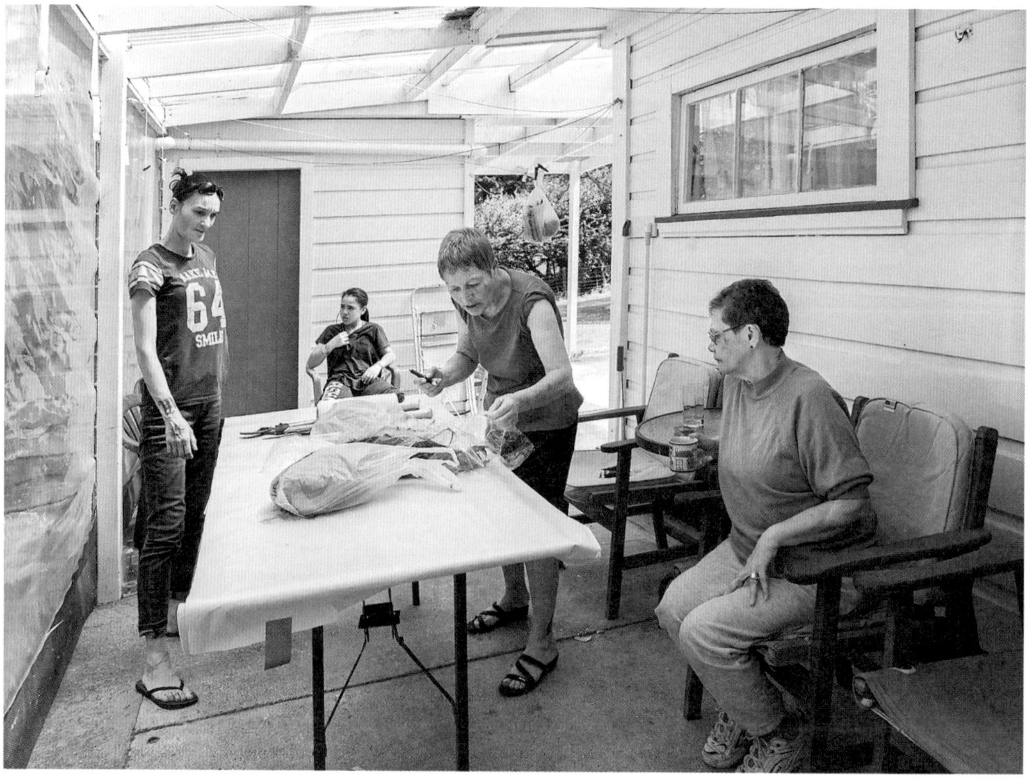
64

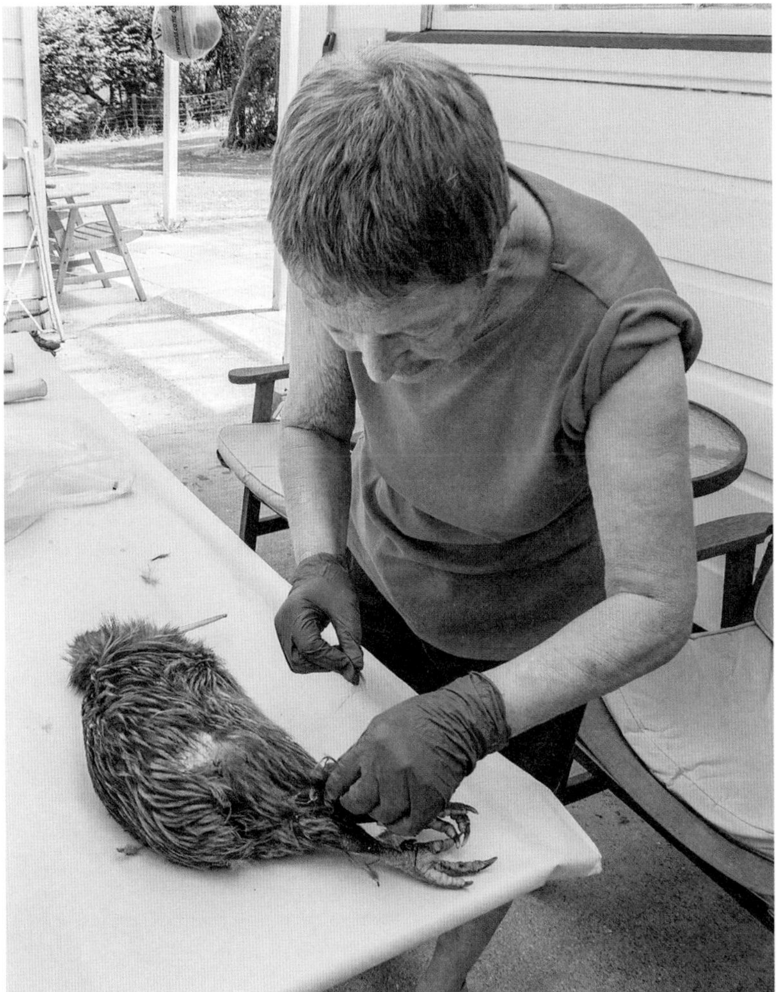

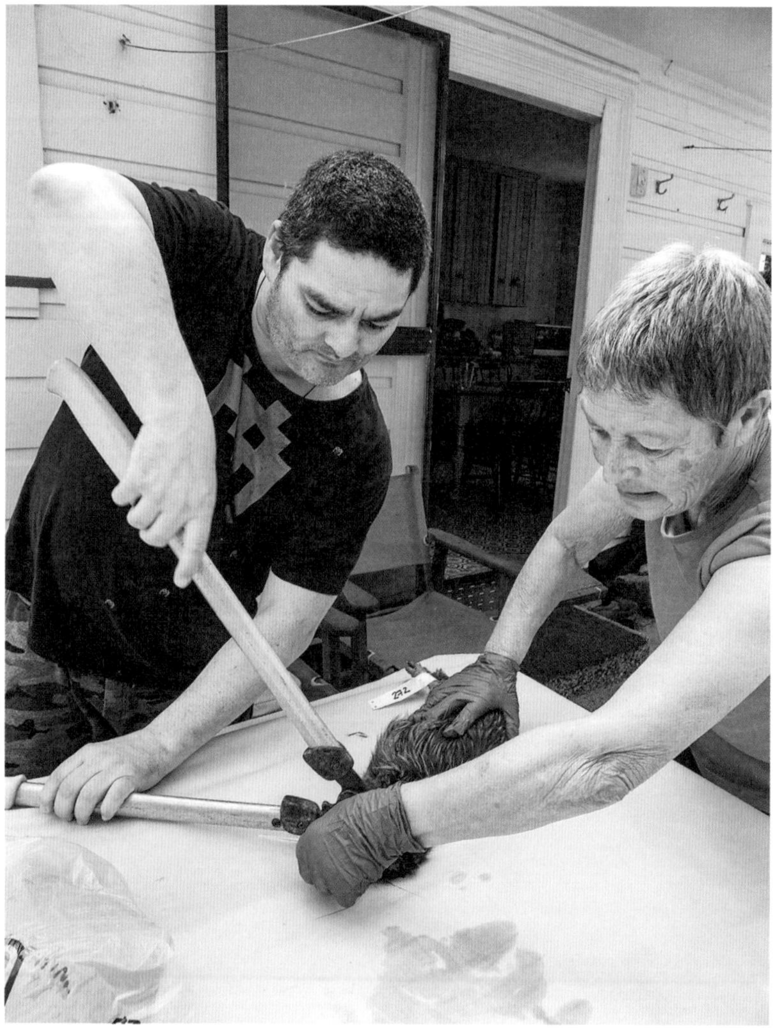

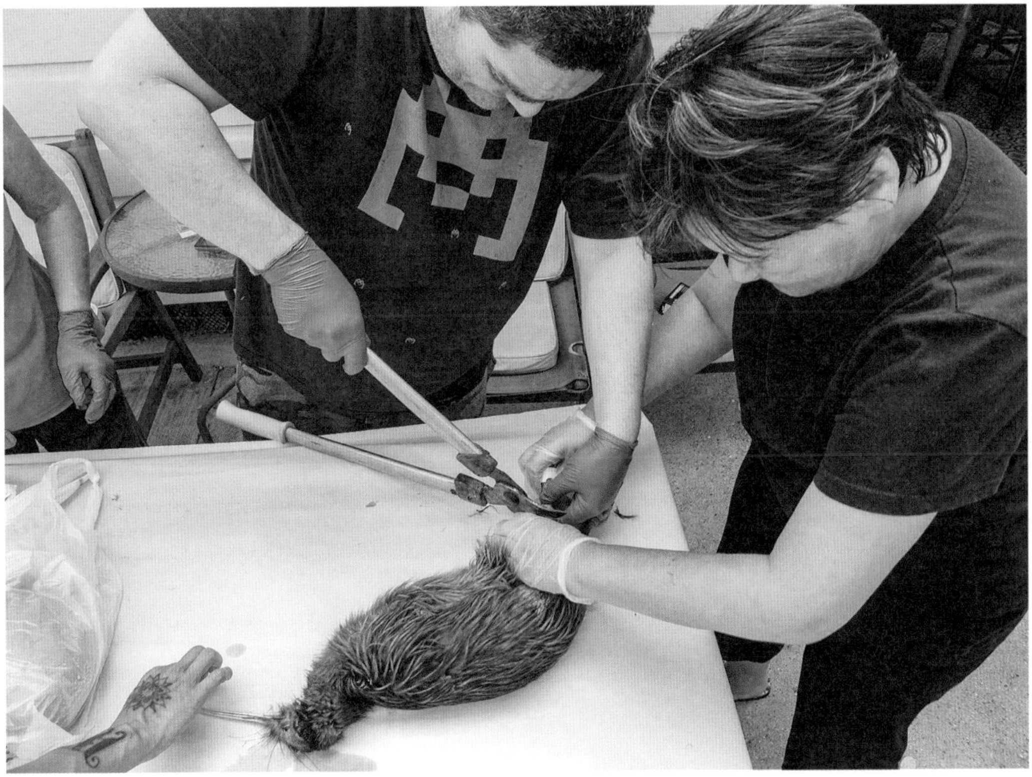

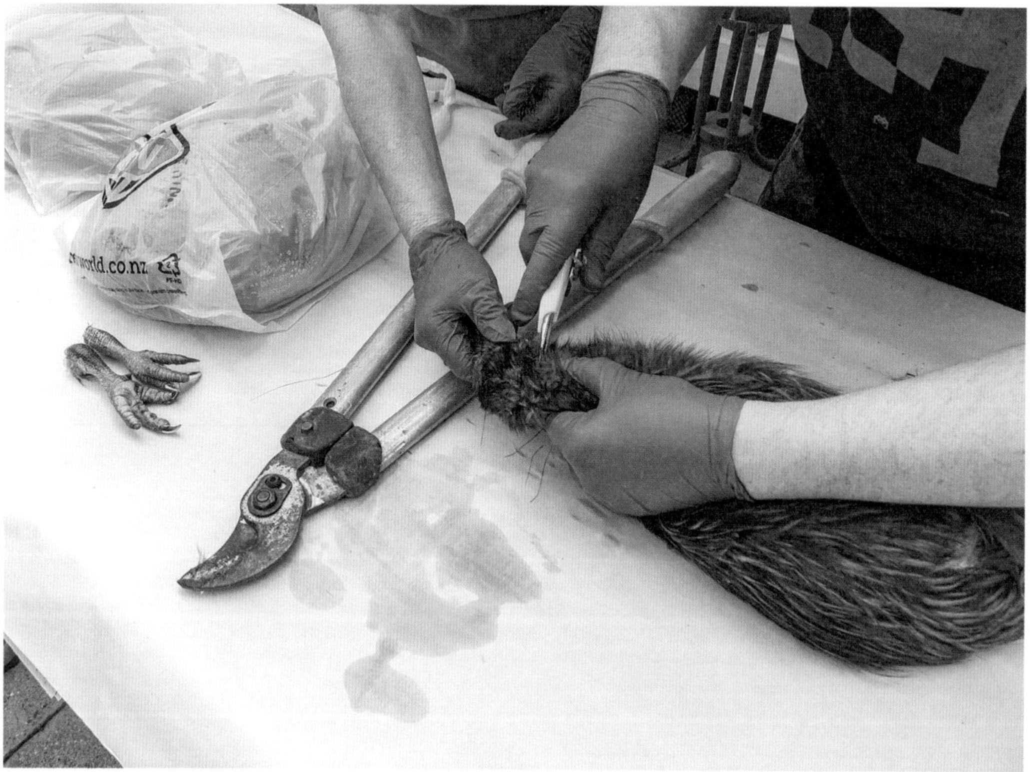

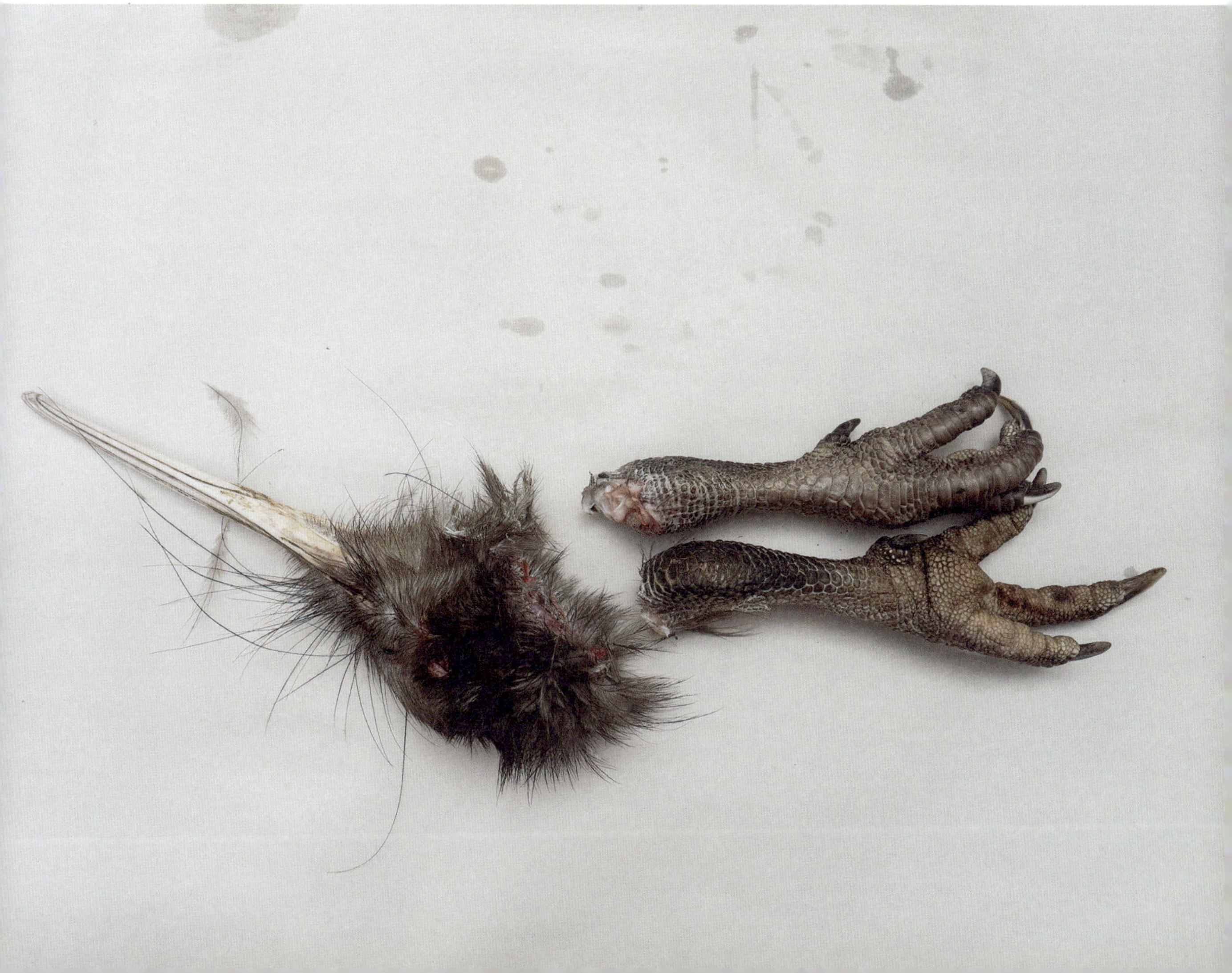

MAHI TAHI HEI HOA TIRITI

Rolien Elliot
Area Manager, Te Papa Atawhai Pēwhairangi Department of Conservation Bay of Islands

You may have heard the saying, 'If you pick a job you love you will never work a day in your life.' I love my work; perhaps that's why I've been in conservation for over 30 years. One of the best things about my job is working with others to protect what is unique to Aotearoa, especially working with whānau, hapū and iwi.

How fortunate we are to have Section 4 of the Conservation Act 1987 that not only encourages the department to work closely with Māori, but also mandates it. Section 4 states: 'This Act shall so be interpreted and administered as to give effect to the principles of the Treaty of Waitangi.'

So, what do these words in Section 4 mean? In using the word 'shall', Parliament has made the obligation under Section 4 mandatory. 'Interpreted', in the context of Section 4, means that any analysis and understanding of the conservation legislation will honour the Treaty principles alongside any conservation objective and determine what is reasonable in particular circumstances. 'Administered' refers to how conservation efforts will be put in place and managed, including a responsibility for the implementation of the law.

Section 4 is central to the Department of Conservation's business. It means that at all levels of the organisation, when carrying out the work the conservation legislation tasks DOC with, the legislation must be interpreted and administered to ensure that the principles of the Treaty will be upheld.[1] Good faith, respect, reciprocity and reasonableness are key.

I believe that DOC can achieve healthy relationships with Māori and meet the Crown's obligations to Māori under the Conservation Act by applying Treaty principles practically in our work. The Treaty principles most relevant to our work include:

Partnership The Crown and Māori must act towards each other reasonably and in good faith. These mutual obligations of cooperation are at the core of the Treaty partnership.

Informed decision-making Both the Crown and Māori need to be well informed of the other's interests and views. When exercising the right to govern, Crown decision-makers need to be fully informed. For Māori, full information needs to be provided in order to contribute to the decision-making process. Consultation in good faith leads to well-informed decision-making.

Active protection The Crown must actively protect Māori interests retained under the Treaty, including the promise to protect tino rangatiratanga and taonga. Active protection requires informed decision-making and judgement to determine what is reasonable in particular circumstances.

Redress and reconciliation The Treaty relationship should include processes to address differences of view between the Crown and Māori. The Crown must preserve capacity to provide redress for proven grievances. Māori and the Crown should demonstrate reconciliation as grievances are addressed.

The special relationship between DOC, as part of the Crown, and Māori includes providing for kaitiakitanga, customary use and mātauranga Māori. One of DOC's ten-year 'stretch goals' is that 'Whānau, hapū and iwi are able to practise their responsibilities as kaitiaki of natural and cultural resources on public conservation lands and waters.'[2] This is in response to the Waitangi Tribunal report on Wai 262,[3] the fauna and flora claim:

> Māori place enormous value on the conservation estate, at two levels. First, it is not only a vast landscape by New Zealand standards; it is also where most of the surviving taonga places can be found. Unlike the rest of New Zealand, which has been so heavily modified by farming, urbanisation, and other land use changes, many parts of the DOC estate remain similar, at least, to that in which te ao Māori was created. And although it is owned by the Treaty partner, every inch of it is tribal territory. Landscapes and landforms evoke the old stories, and they in turn evoke whakapapa. For this reason, individual iwi and hapū relationships with conservation land remain tangible in ways not usually possible in more modified environments. Sometimes relationships with flora and fauna can be recreated within the DOC estate as substitutes for those that have been lost elsewhere.
>
> Secondly, DOC is responsible for almost all remaining indigenous flora and fauna species — many of which are found nowhere else in the world, and many of which are threatened and/or endangered. For most iwi and hapū, the department controls access to and relationships with such taonga. Without them, mātauranga Māori simply could not survive.[4]

Relationships between the Crown and Māori can be complex, particularly in a pre-Treaty-settlement environment like Northland, with overlapping areas of interest. Some whānau, hapū and iwi carry the grievances of multiple generations and do not trust the Crown, or DOC as part of the Crown, and do not wish to engage. I acknowledge this grievance, but I am saddened by the lost opportunities. We might not be able to make up for the last 200 years but we can work proactively in the spirit of Te Tiriti o Waitangi to do whatever is possible to meet our obligations going forward, and hopefully create opportunities to work together and learn from one another.

To deliver DOC's obligations under Section 4 of the Conservation Act at a local level, my strategy has been to build effective relationships based on shared understanding, trust and respect with whānau, hapū and iwi on the ground. Providing opportunities to meet on-site allows connection to the whenua and encourages storytelling and whakapapa to emerge. The sharing of information and knowledge, and working together on issues and solutions, builds rapport and an understanding of each other's perspectives. Often, we are on the same kaupapa, so it makes sense to work together.

Being a female DOC manager can sometimes bring challenges in engagement with Māori; often being on the back bench at marae, without the opportunity to speak (depending on the tikanga in place) can make it difficult to acknowledge the kōrero and communicate messages that need to be heard. Fortunately I have people I work with who can support me on the marae and after hui. Conversations often begin in the wharekai, with the wāhine, whaea and kuia talking about what needs to happen and who are the best women for the job!

DOC, as a government agency, must work within our legislative frameworks as well as meet our Treaty responsibilities. Conservation management involves a wide range of decision-making, from day-to-day operations to consents, licences and permits. While iwi do not have final decision-making power, their involvement in the process is critical and provides strong guidance. For example, kiwi are absolutely protected by the Wildlife Act, dead or alive. Holding or using manu or huruhuru for cultural purposes, such as making kākahu, requires a permit from DOC. One could say Māori have a right to this taonga, and the requirement to apply for a permit may be seen as an insult. While I appreciate this view, we still need to operate within the law. By working with hapū we have developed a set of criteria that is used when we assess permit applications. Does the person wishing to hold or use the huruhuru manu have the support of the hapū from where the taonga came? Is the use appropriate and for cultural purposes, not for commercial gain?

The permit process has advantages. It helps keep track of the use of cultural materials and ensures the use of protected taonga is appropriate. For example, no one would want to see the mana of kiwi degraded by their feathers being used in souvenirs sold to tourists for individual gain.

I remember asking myself, after three dead kiwi had been brought into the office in as many weeks, how we could get the message out that people need to control their dogs and slow down on roads where kiwi live nearby, and what we should do with all the dead kiwi in our freezers. I shared my thoughts with Ngāti Torehina ki Matakā kaumātua Hugh Rihari when he visited the office. We took a photo of ourselves surrounded by dead kiwi that were being stored in the freezers. The photo certainly told a story, and the local newspapers published it with some key messages about protecting kiwi. There was good feedback on the article, but that didn't help with the kiwi in our freezers.

Some time after this, Hugh's wife, Raewyn, rang me to arrange a time to come and see me; she and Tiwai Rihari-Rawiri had a proposal for me. They are smart and generous, those whaea. Raewyn and Tiwai's offer was to run wānanga huruhuru manu as a way to utilise the manu and to teach wāhine the cultural practice of pelting and use of feathers in kākahu, with DOC helping with the venue and equipment. What an awesome idea.

Local hapū in kiwi areas were asked to nominate weavers who could be permit-holders for the manu and feathers. With a permit-holder in place, native birds could be logged and measured then returned directly to the hapū from the rohe they were found in, and would not end up in DOC freezers.

Although sad to have dealt with dead kiwi, I am pleased that the mana of these birds will carry on as treasured kākahu. The wānanga huruhuru manu were a great success. The participants were enthusiastic and engaged in the traditional methods of pelting, and through the experience and kōrero they have become kiwi protection advocates in their communities.

The wānanga huruhuru manu is a good example of collaboration, with Māori women and DOC women working together to solve a problem and exercise mātauranga Māori. Effective partnerships between DOC and tangata whenua are of immense value and can enhance the preservation of our natural, historical and cultural heritage. I look forward to the next stage of engagement with whānau, hapū and iwi, one of co-design and co-management.

These are my personal views, not necessarily the views of the Department of Conservation.

1 'Peka 4: The Principles of the Treaty of Waitangi', Part 4 of 'Te Kete a Te Rito', the training manual of Te Pukenga Atawhai (Wellington: Department of Conservation, August 1999): 65–76. Te Pukenga Atawhai is a training programme undertaken by all DOC staff, which covers the skills needed to build and maintain effective relationships with tangata whenua.

2 'Department of Conservation Statement of Intent 2016–2020', May 2016, 10, https://www.doc.govt.nz/Documents/about-doc/role/publications/statement-of-intent-2016-2020.pdf.

3 The Waitangi Tribunal is a permanent commission of inquiry established in 1975 under the Treaty of Waitangi Act to consider and make recommendations on claims brought by Māori about Crown acts or omissions that breach the promises made in the Treaty. Wai 262 is the 262nd claim registered with the Waitangi Tribunal. It was lodged in 1991. The claim calls for the protection of Māori culture, identity and traditional knowledge in New Zealand's laws and in government policies and practices, as well as a recognition of the place of core Māori cultural values in contemporary New Zealand life, such as iwi and hapū acting as kaitiaki of taonga, traditional knowledge, artistic and cultural works, important places, and flora and fauna that are significant to iwi or hapū identity. In general, the tribunal's findings and recommendations regarding Wai 262 concern the contemporary relationship between the Crown and Māori rather than addressing past grievances.'Ko Aotearoa Tēnei: Report on the Wai 262 Claim Released', accessed 11 September 2019, https://www.waitangitribunal.govt.nz/news/ko-aotearoa-tenei-report-on-the-wai-262-claim-released.

4 'Ko Aotearoa Tēnei, Te Taumata Tuatahi: Claims Concerning New Zealand Land and Policy Affecting Māori Culture and Identity', Wai 262, Waitangi Tribunal (Wellington: Legislation Direct, 2011): 127.

THE NORTHLAND BROWN KIWI

Cinzia Vestena
Biodiversity Assets Ranger, Te Papa Atawhai Pēwhairangi
Department of Conservation Bay of Islands

The Northland brown kiwi is one of four distinct brown kiwi taxa in New Zealand. Their current distribution extends from Whakaangi Landcare Trust on the Hihi Peninsula, Doubtless Bay, in the north, to the translocated population at Tāwharanui Open Sanctuary (an hour north of Auckland) in the south. Northland brown kiwi are present on some offshore islands but they are all but extinct from southern Northland. BirdLife International (2018) considers that the IUCN (International Union for Conservation of Nature) Red List category for brown kiwi is 'Vulnerable', with a decreasing population trend considered to be facing a high risk of extinction in the wild.[1]

Northland is considered to be a stronghold for brown kiwi thanks to the widespread community efforts in predator control. These include collaborative projects involving landowners, the Department of Conservation, Northland Regional Council, landcare groups, iwi and forestry companies. Often neighbouring landowners have combined their protection efforts to maximise benefits to kiwi, with some large projects forming charitable trusts or incorporated societies. Currently, most managed kiwi populations where predator control is in place are growing at rates of two per cent per annum or higher. However, the majority of kiwi are found in unmanaged sites and are still in decline.

Current threats to Northland brown kiwi are habitat loss and predation by introduced mammals, especially dogs, ferrets, stoats and cats. Dogs and ferrets are known to kill adults. The loss of any adult greatly reduces population potential in long-lived species like kiwi. In Northland, kiwi are also vulnerable to vehicle strikes, capture in possum traps set on the ground, and falling into water troughs.

DOC's Kiwi Recovery Plan 2018–2028 identifies dogs as the greatest threat to adult kiwi in areas where kiwi habitat overlaps with, or is close to, human populations.[2] Any dog can kill a kiwi, even small or soft-mouthed dogs. Kiwi lack wings, feathers and muscle in the critical parts of their body, which means that the chest of a kiwi is easily crushed in a dog's mouth. Even a dog mouthing a kiwi over its back can crush vital organs and cause internal bleeding. This issue is most pronounced in Northland, where dog kills are the main reason that the average kiwi life expectancy is only 14 years compared with a life expectancy of 30–40 years elsewhere in the North Island.

Dogs of all sizes, breeds and training are attracted to the smell of kiwi. The predation events observed with dogs do not fit with traditional predator–prey models because dogs can roam long distances and kill far more prey than is needed to sustain them. Kiwi are particularly vulnerable near human settlements; dogs kill kiwi both at night and during the day when they may be sheltering beneath dense vegetation along the edges of roads and tracks.[3]

The most dramatic example of the potential scale of destruction by dogs was in 1987 at Waitangi Forest in the Bay of Islands, where a single dog is believed to have killed 500 kiwi over a six-week period.[4] This was not an isolated incident. Dogs were responsible for 135 (70 per cent) of the reported kiwi deaths in Northland between 1990 and 1995.[5] A study in central Northland from 1994 to 2008 recorded dogs being responsible for 22 of the 44 adult deaths of known cause.[6] Dogs were also the leading cause of death of sub-adult kiwi (37 per cent of the deaths from a known cause). There is also recent evidence that dogs have a significant impact on kiwi in the wider Kerikeri area. In 2015, two dogs hunting together killed at least eight kiwi over a few months. More recently, six dead kiwi found at one site in the Bay of Islands area were confirmed to have been killed by dogs.

Kiwi chicks suffer exceptionally high mortality rates in unmanaged sites, especially in the first 100 days of life. Chicks forage independently and have no behaviours or defences against predators.[7] Cats will roam up to 20 kilometres from home, and they hunt at night, when kiwi are active. In unmanaged sites in central Northland 94 per cent of chicks fail to reach adulthood due mainly to stoat predation, followed by cat predation.[8]

As part of the kiwi conservation effort, DOC expects any dead kiwi, or any other protected wildlife, to be handed in to the closest office. Once the kiwi are received, they are examined to try to identify the likely cause of death. When it is suspected a dog may have killed the kiwi, we send the bird to Massey University in Palmerston North for a post mortem to ascertain if that was the case.

Keeping records of the numbers of dead kiwi and what caused their deaths helps us identify risks. For example, between January 2018 and September 2019 we received 39 dead kiwi: four died of natural causes, eight were cause unknown, eight were killed by dogs, one drowned, one was predated by cat or stoat, one was caught in a possum trap set on the ground, and 16 were hit by cars.

Every incident gives the department an opportunity to engage with the person who found the dead kiwi and advocate for better protection, such as setting up possum traps off the ground and making pools or cattle troughs kiwi-proof in very dry summers. Vehicle strikes mainly occur where kiwi are in high numbers near a cluster of houses. Road signs set up by local residents asking drivers to slow down at night for kiwi are a very common sight in the Bay of Islands.

Another, equally important, reason for any dead kiwi to be handed in to DOC is that it allows us to return the kiwi to the hapū or marae of the rohe where it was found so that its feathers can be given a second life through cultural traditions. Kiwi, and other manu, are temporarily kept in freezers at the local office until there is a request to use their feathers.

•

Kiwi distribution maps were first drawn in 1996 to identify priority management areas for kiwi recovery. By identifying areas of high kiwi density, we can advocate for long-term kiwi recovery and statutory provisions can be put in place to manage the threats of dogs and cats.

The maps identify areas of high kiwi density and 'kiwi present' areas based on annual kiwi call monitoring, which has been the accepted methodology to determine kiwi presence, abundance and distribution since 1993. The number of kiwi calls fluctuates each year due to natural events or management. Annual call counts must take place for at least five years, but preferably 10 years or more, to track long-term changes or trends. The maps are reviewed every five years.

Since 1998 DOC has advocated for 'no cats and dogs' consent conditions for proposed land development in high-density kiwi areas (where there is an average of five calls per hour or more). In 'kiwi present' areas (with an average of fewer than five calls per hour) landowners are encouraged to minimise the potential risk their cats and dogs may pose to kiwi by being responsible pet owners.

While the maps are a useful tool, they cannot be completely accurate. Reports of kiwi heard or seen in areas where they were not thought to be present are fast becoming more common thanks to better communications and awareness within communities, alongside active predator control helping kiwi to breed successfully. It is not possible to state with certainty that kiwi are not present outside the identified areas as they are nocturnal and because not all birds call. They also may have such low numbers in a particular area that it is difficult to detect their presence.

Despite the comprehensive data provided by the annual call counts carried out at multiple listening sites in Northland, populations are dynamic, with juvenile kiwi moving to find suitable territories, and the boundaries of the different density areas are never exact.

It is important to note that kiwi will live almost anywhere and are known to travel to find suitable habitats. They don't need native forest and are often found in scrub, farmland, sand dunes and among exotic plants and mangroves. They particularly like wetland vegetation and tree-covered areas close to rivers.

Since 1993, the kiwi call count has been taking place annually. Over the years, many listening stations have been added, predominantly in areas where community groups are working to protect kiwi. During the 2017 kiwi listening period, counts took place at 24 permanent stations in eastern Northland, where the kiwi shown in this book were found, with an average of 16.7 calls per hour. Over the years there have been many natural fluctuations but overall there has been an upward trend in the number of calls heard, with the eastern area results consistently the highest in Northland.[9] High kiwi call rates don't necessarily indicate low risk of a population crash, as call rates can decline rapidly. For example at Katui, near Trounson Kauri Park, call rates dropped from 40 per hour to zero in just a few years.[10]

Kiwi call to maintain territories, but also to preserve the pair bond. Pairs and breeding kiwi often duet, with birds responding to their partner's call, and newly paired kiwi call more frequently than well-established pairs. Most kiwi call only occasionally and can be heard up to two kilometres away in ideal listening conditions. Males are heard approximately three times more often than females — males tend to call more often, and their sharper-sounding call carries further than the lower-pitched calls of females. Juveniles don't generally start calling until they are 18–24 months old.

Kiwi call all year round, but the peak in calling usually coincides with mating and the start of each incubation period (April–June in Northland).[11] The annual call count is done at the same time each year to take advantage of the optimum conditions for kiwi calling during the darker moon phases, as there is evidence of female kiwi calling less frequently on moonlit nights at some sites. Listening is for two consecutive hours on four different nights (not necessarily in a row, as listening should not take place on rainy or windy nights) and starts no earlier than 30 minutes after sunset when kiwi start to emerge from their burrows. Peak calling usually occurs in the first half of the night for both sexes. They continue to call sporadically until sunrise when they return to their burrows to sleep.[12]

•

The recovery and protection of Northland brown kiwi relies on the engagement and collaboration of many stakeholders: tangata whenua, local and national government agencies, and community and landowner-led kiwi conservation projects. The long-term success of these projects requires ongoing energy and enthusiasm on the part of the groups and individuals involved, to control pets and advocate for responsible dog and cat ownership, as well as ongoing funding to sustain protection efforts. Despite the wider community's best efforts, there will always be loss of manu, but collaboration between local DOC staff and tangata whenua will ensure that these taonga remain cherished in the future.

The wānanga huruhuru manu provided a great opportunity for DOC staff to learn about how Māori see the kiwi being returned to life by using their feathers to make traditional ceremonial garments. Having to examine dead kiwi for possible cause of death is sad, especially when handling beautiful adult kiwi that could have been continuing to breed. The idea that the kiwi is given a second chance at life is uplifting.

My role in the wānanga was to organise the work space at the office, including having the right tools, and to ensure all manu were correctly labelled with their rohe so that they could be returned to the correct hapū. Under the Wildlife Act 1953 the manu must be accompanied by an authority to hold protected wildlife (or part of it, such as feathers). This can only be issued by DOC to a kaitiaki supported by their hapū or marae.

For me it was a privilege to be involved in the wānanga, to be part of the community coming together and sharing and learning.

---

1 H. Robertson, personal communication, 24 September 2018.

2 J. Germano, S. Barlow, I. Castro, R. Colbourne, M. Cox, C. Gilles, … S. Yong, 'Kiwi Recovery Plan 2018–2028' (Wellington: Department of Conservation, 2018).

3 R. J. Pierce and W. Sporle, 'Causes of Kiwi Mortality in Northland', Conservation Advisory Science Notes, No. 169 (Wellington: Department of Conservation, 1997).

4 M. Taborsky, 'Kiwis and Dog Predation: Observations in Waitangi State Forest', *Notornis* 35 (1988): 197–202.

5 Pierce and Sporle, 'Causes of Kiwi Mortality in Northland'.

6 H. A. Robertson, R. M. Colbourne, P. J. Graham, P. J. Miller, and R. J. Pierce, 'Experimental Management of Brown Kiwi (Apteryx mantelli) in Central Northland', *Bird Conservation International* 21 (2011): 207–220.

7 J. A. McLennan, M. A. Potter, H. A. Robertson, G. C. Wake, R. Colbourne, L. Dew, … J. Reid, 'Role of Predation in the Decline of Kiwi, Apterix spp.', *New Zealand Journal of Ecology* 20 (1996): 27–35.

8 Robertson et al., 'Experimental Management of Brown Kiwi'.

9 E. Craig, 'Call Count Monitoring of Northland Brown Kiwi' (Whangarei: Department of Conservation, 2017).

10 H. Robertson, personal communication, 24 September 2018.

11 Germano et al., 'Kiwi Recovery Plan 2018–2028'.

12 H. Robertson, personal communication, 24 September 2018.

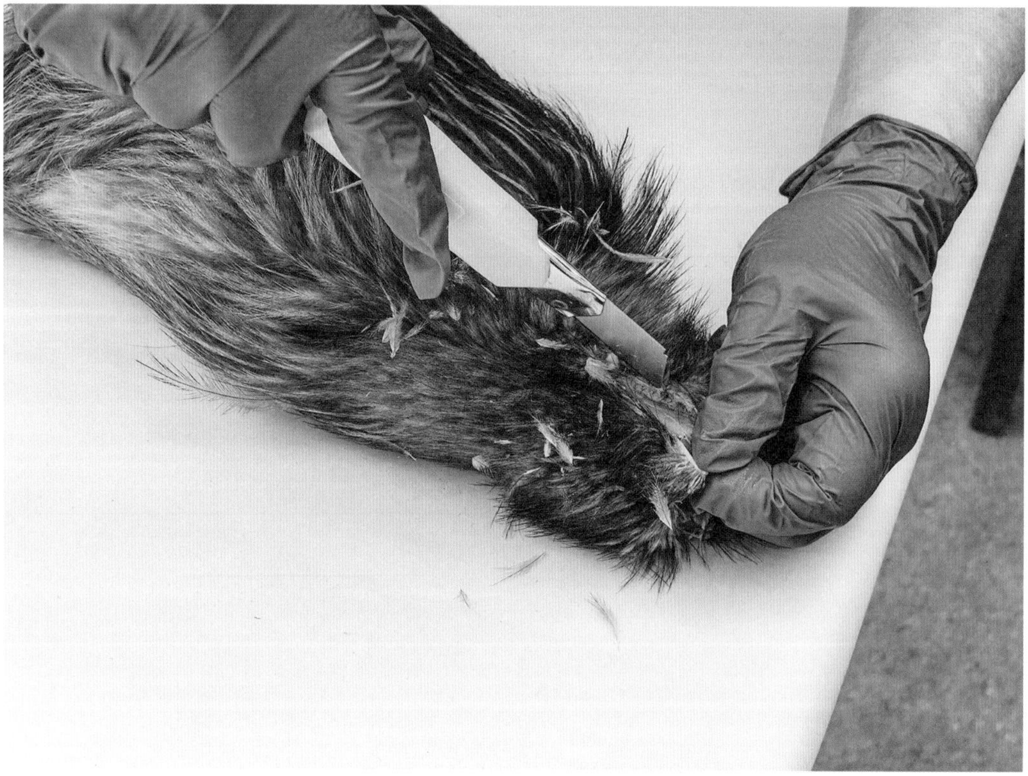

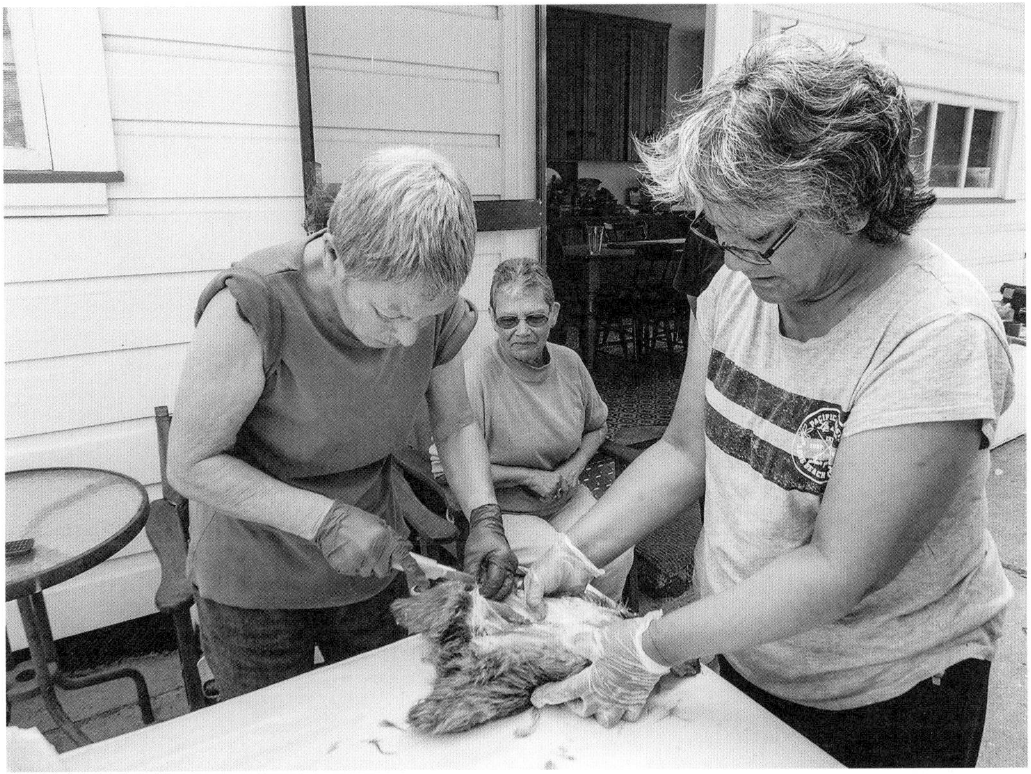

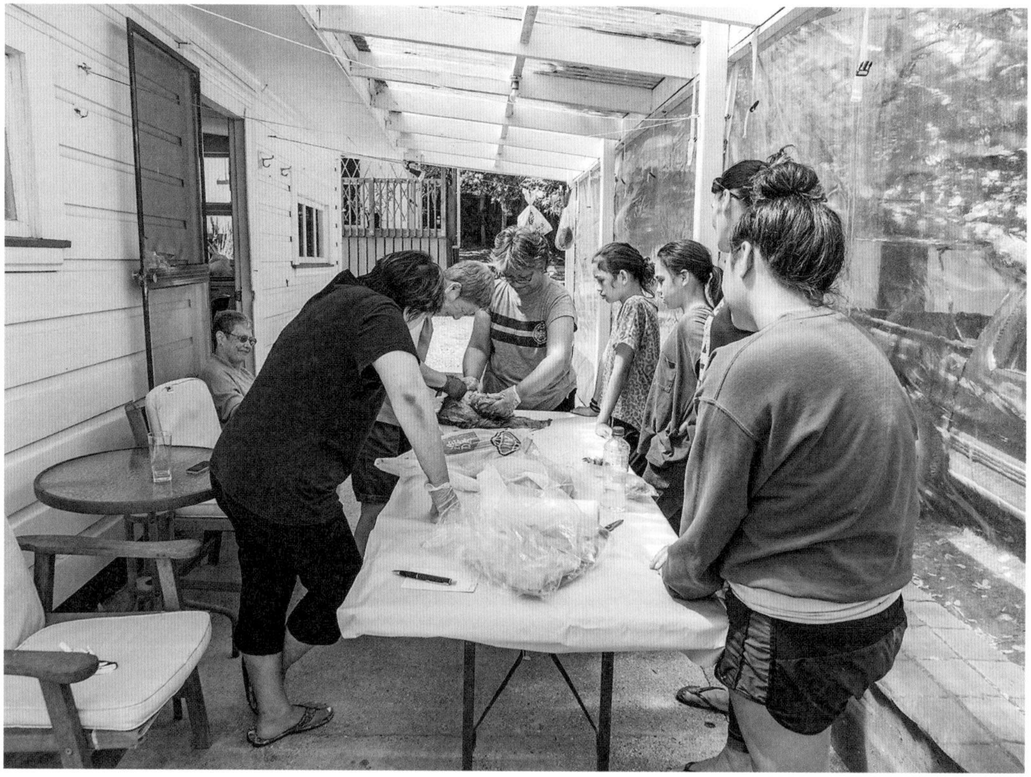

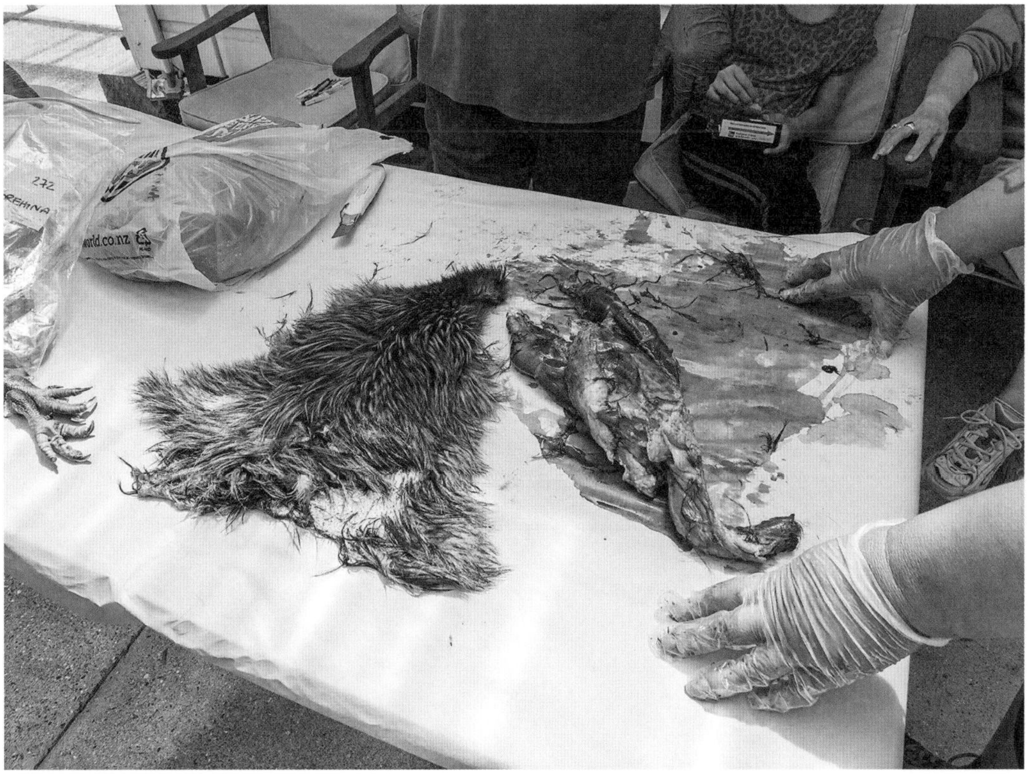
272
world.co.nz

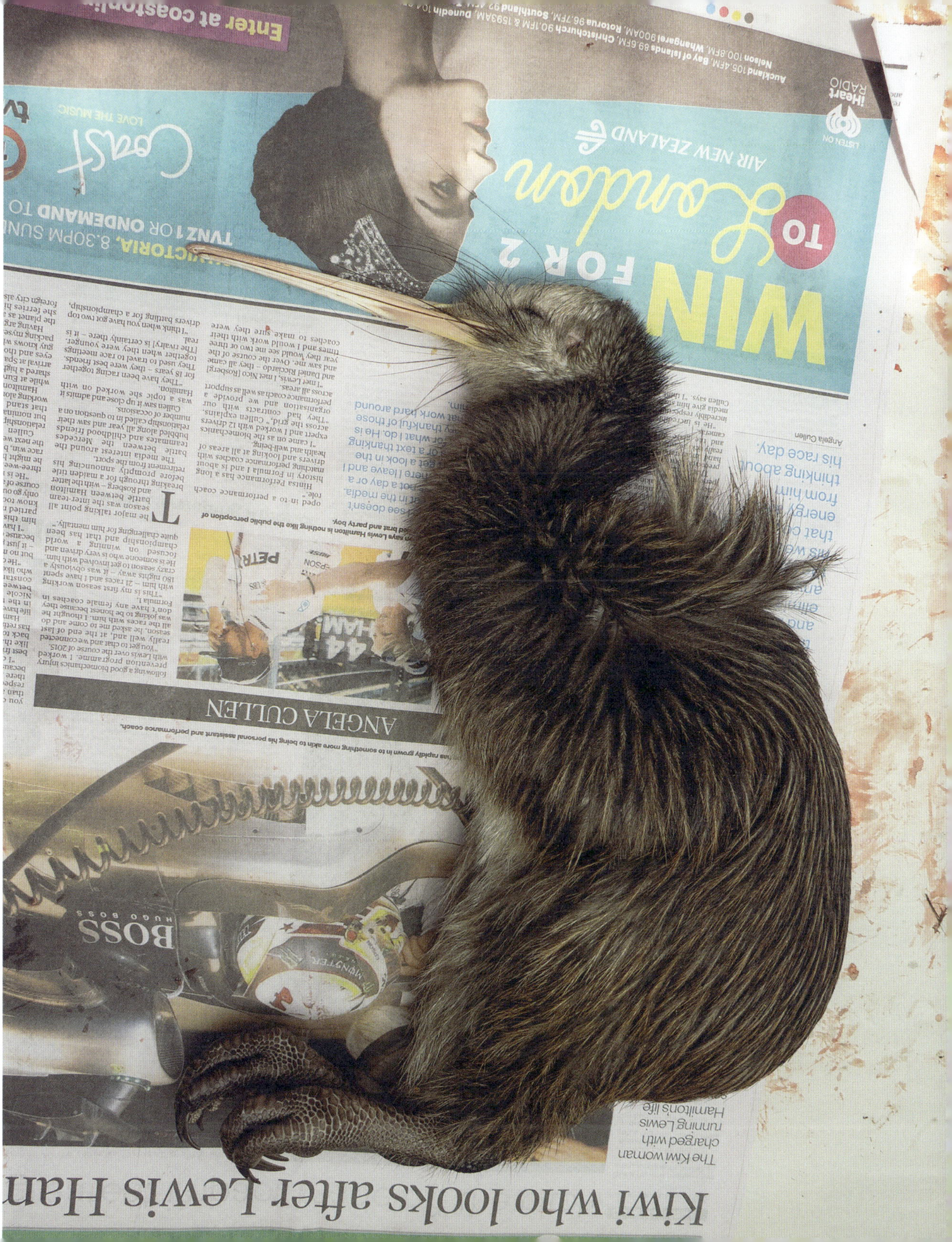
Kiwi who looks after Lewis Ham
The Kiwi woman charged with running Lewis Hamilton's life
ANGELA CULLEN
WIN FOR 2 TO London
AIR NEW ZEALAND
Enter at coastonli
BOSS

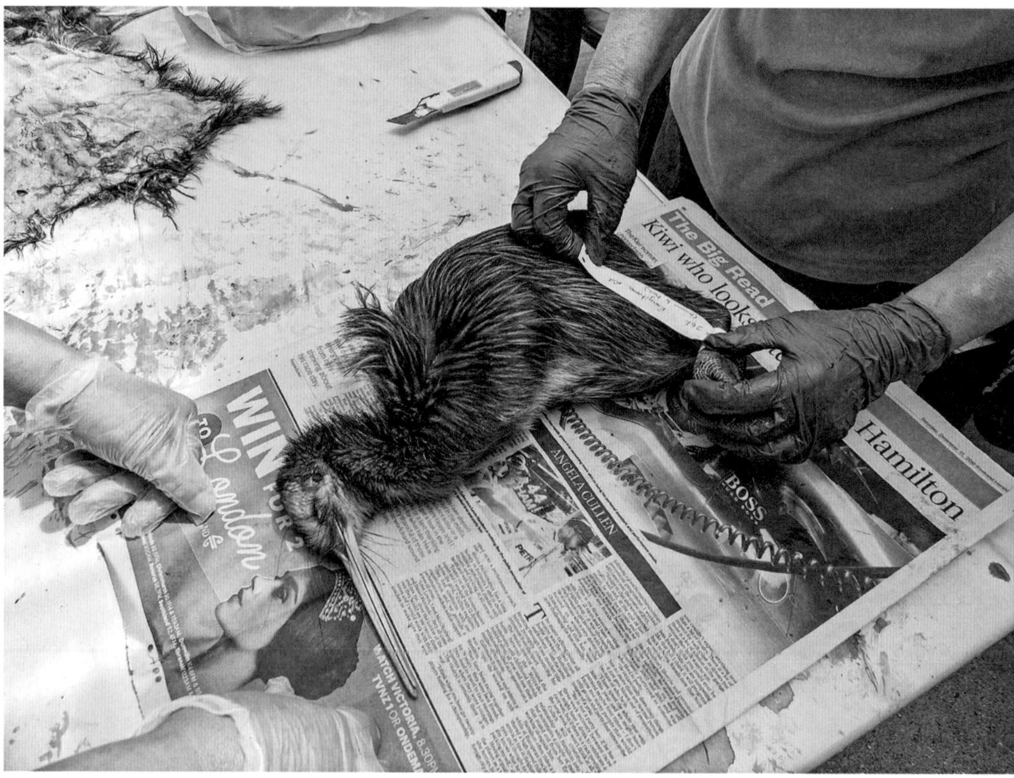

The Big Read
Kiwi who looks
Hamilton
ANGELA CULLEN
BOSS
WIN

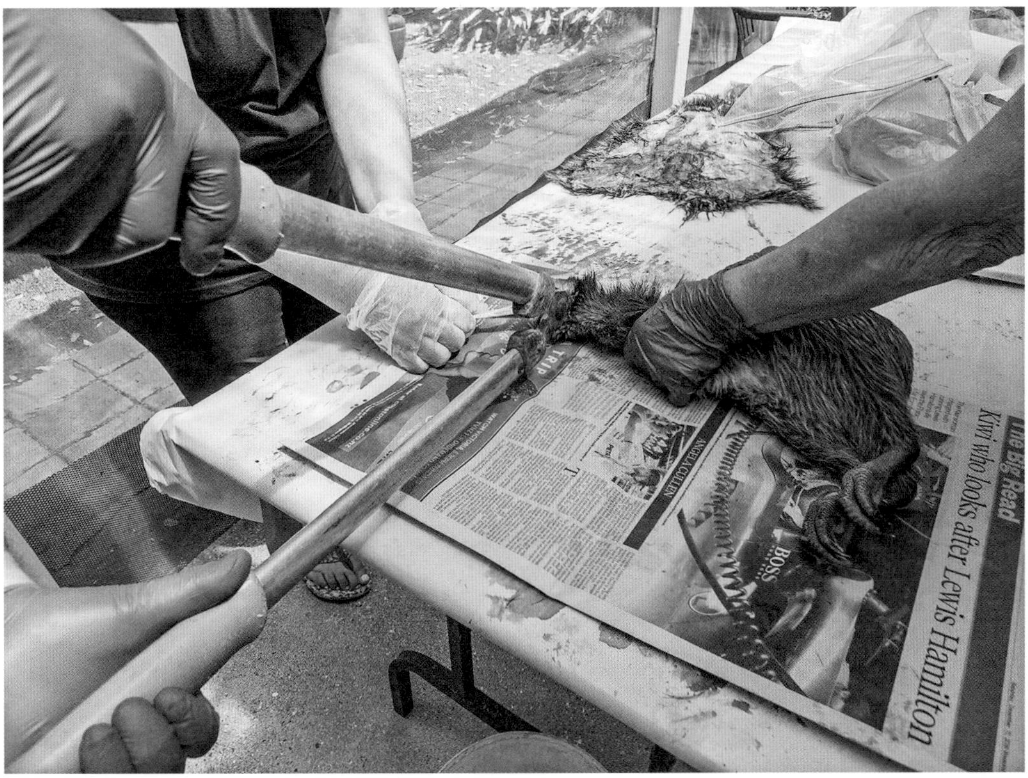
The Big Read
Kiwi who looks after Lewis Hamilton
TRIP
BOSS

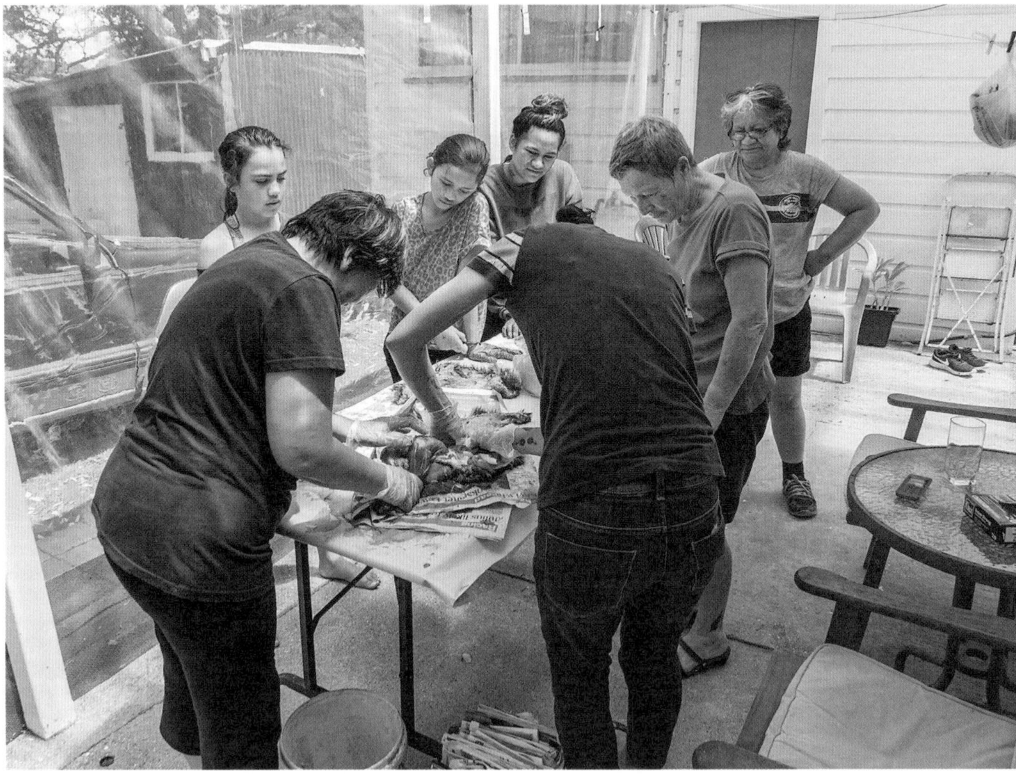

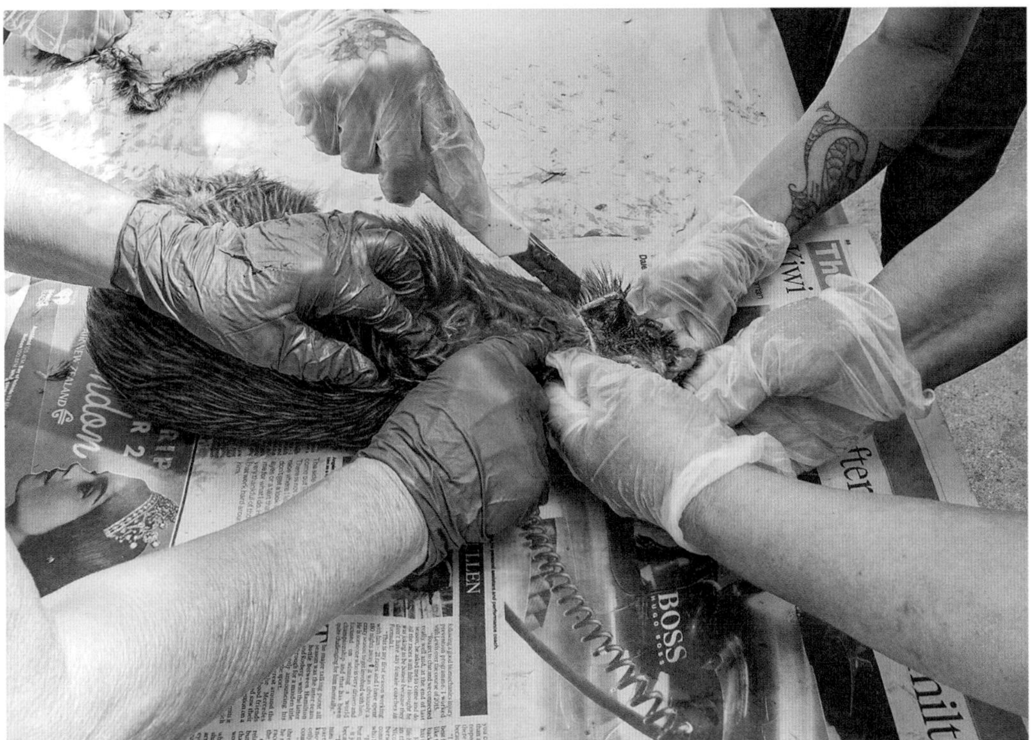
Kiwi
BOSS

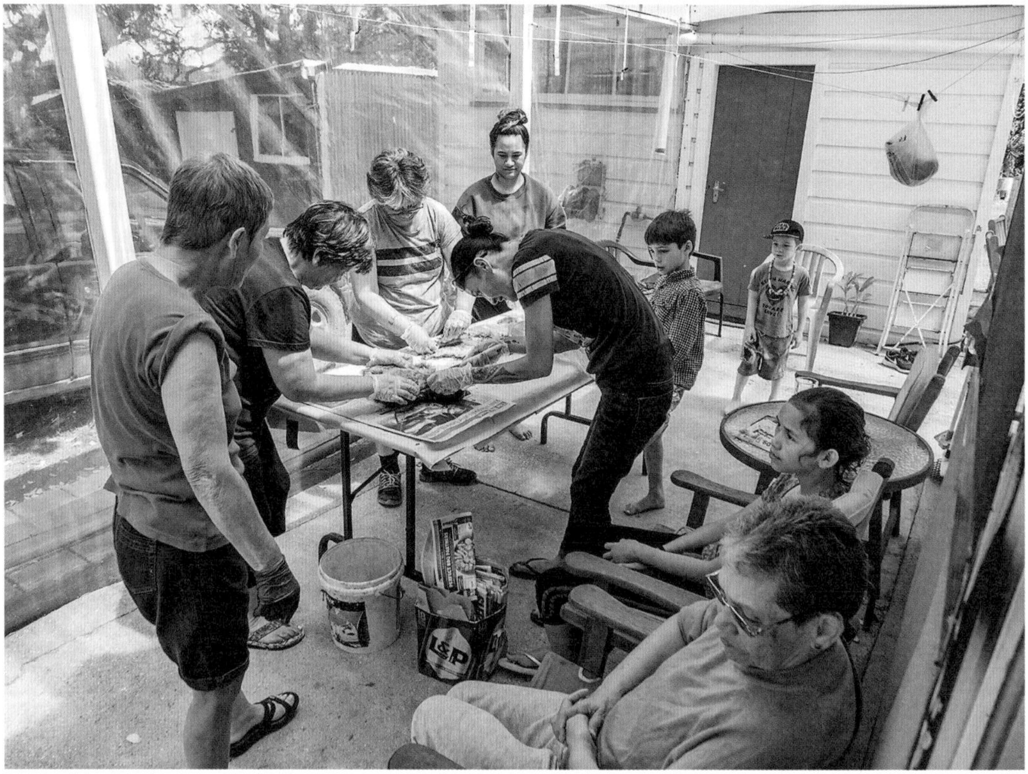
L&P

INTO THE VOID

Matariki Williams Ngāi Tūhoe, Ngāti Hauiti, Taranaki, Ngāti Whakaue

There is a photo on my sideboard. It is of my great-grandfather Kupai McGarvey, Dad's Koro. He wears a suit and tie, over the top of which is tied a kākahu. The kākahu has a border of feathers running along the top in a row of white, a row of black, and then white again. The body of the kākahu, from what I can see, has columns of more feathers interspersed with hukahuka of wool. This is a photograph that has held a place on the walls of many of my whānau members' houses. From childhood I've looked in awe at my koroua, who passed before I was born, as he stands in a kākahu I have never seen, holding the tokotoko that maintains a rarefied presence in our whānau.

Today I looked more closely at the kākahu. Informed by recently acquired knowledge from my role as a curator with privileged access to the taonga Māori collection in our national museum, I realised that the kākahu worn by my koroua is not made from muka. It is apparent that the kaupapa of the kākahu is made from canvas, though its golden colouring is evocative of some of the muka that I have seen in the taonga Māori collection. Muka itself, the processed form of certain types of harakeke, has a sheen to it and a silkiness that invites touch — in museum collections it is a touch that is mediated by rubber gloves. Not really a touch at all, and not really muka at all, but there's no shame in that.

•

To these arresting photographs of the wānanga huruhuru manu. The opening composition is one of tranquillity, reminiscent of the late afternoons of a holiday, with a lake or beach in the background, lapping sand softly. To the right of the photograph the sun beckons to the water, *come for your final swim of the day.* The water is surrounded by a dense border of native trees, a kāinga for our native species, and a hiding place for their prey. Looking closely again, the house and its external objects have a familiarity that I can recognise, of a home where there is always a seat to sit in. At houses like this there are projects, objects in varying stages of process. They are the houses that exist on all sides of my whānau. Like the porch of my Koro Jim, Mum's dad, where rīwai and paukena, the bounty from his garden, would be waiting to be given to visitors, his seeds drying for next year's garden beds. One year there were his kānga drying too. I never did get to try them, and I wonder what has happened to his garden in the five years since he passed.

In this photo, the first of the series, I make out the silhouette of a kiwi hanging in a plastic bag. Lit by this most peaceful of lights, its suspended body is incongruous with the surrounds. When I point it out to my kids they respond with a mix of horror and intrigue: *Why is it there? Who hurt it? Is it dead? Are they all dead?* This photo, and the others of disembodied kiwi parts in various states of frozen and thawed, conjure yet another memory, this time of the kākā in the freezer at our family home in Tauranga. The body of this bird was recovered from under our house. Dad had found it in a dazed state and contacted DOC, who informed him that the kākā had likely been eating berries that made it a little disorientated so that it had flown into the house and injured itself.

•

We are not a family of weavers. Our hands grip knitting needles not mussel shells. Bonnets, booties, cardigans clothe us, not for us are the kahu kiwi, kahu huruhuru of other whānau. This is true on each side of my whānau. I know no trace of whānau-made or inherited raranga beyond the putiputi I was taught in a primary school noho marae on Matakana Island.

•

I must return to the light of the photo. So filled with potential of the night to come. Of the time when the kiwi awaken. The quietness of that potential belies the brutality of the images to come, yet it is not a brutality that exists without care. The hands that cradle the head as other hands employ large shears to remove it, they work in concert and it is with manaakitanga of the bird, its life and the future use of pelts. They cradle the potential for what the bird's role will now be. It is no coincidence that Te Pō in which the kiwi awakens, *the big night, the long night, the dark night,* is synonymous with potential.

•

As the skinning takes place, butcher's paper is laid on a table and spread with newspapers. Cups of tea and water for everyone. The familiarity of the paper and cups of tea is comforting, their ubiquity in Māori settings, on wharekai tables, out the back of wharekai where pigs are hung and carved, and vegetables peeled and chopped. But this isn't anywhere, this is the Far North. When my Nan prepared to leave Rūātoki for Queen Victoria School, her uncles told her to befriend girls from the North because they would have the 'old reo'.

In the photo, Nan's father, Koro Kupai, holds a tokotoko carved by a man from the North, his name is Moka Puru. The tokotoko was carved during the 1975 Māori Land March, which my koroua joined at Parewahawaha Marae in Bulls. He continued on with Ngā Roopu o te Matakite to Wellington to present their petition.

'Not one more acre,' they chanted, carrying with them a pouwhenua, also carved by Moka Puru, which will not touch the ground until all Māori land has been returned.

The tokotoko remains, but where is the kākahu?

•

'Kiwi who looks after' reads a truncated newspaper headline as six hands reach in to work on the body of a kiwi. By an accident of arrangement, a hand sits at the end of a newspaper advertisement featuring a coiled cord, giving the impression that the kiwi is being operated on.

And so this kiwi will continue to live in the taonga its feathers will adorn.

•

As the kiwi is worked on, the images change. Colour is removed as blood runs out. I've never seen a bird pelt before, and this is not what I imagined. Many times I've seen pigs, cows, deer hanging from the strut of a garage roof or outside a wharekai, their innards removed onto the concrete floor — easier for their blood to be hosed away. Yet there is something in the body of the kiwi, denuded of its feathers, that feels so vulnerable. So inviting for me to cloak in my arms.

•

My mother was not a weaver. Nor was her mother. Beyond that, I don't know. What my mother was, was a knitter; what she gave us was her love manifest in wool. What she passed down was her knowledge of knitting, teaching her right-handed daughters to knit in her left-handed style. Wanting to learn raranga and whatu is not a reclamation as we didn't have the skills in the first place. The kākahu in our family are machine-made, the whenu are cotton and the dyes synthetic; this is what most whānau have access to these days. Where the making knowledge is missing, these newer kākahu have filled the void of potential.

•

A recent stay near the kiwi habitat in the Orongorongo Valley led to a night of hearing kiwi call. At one point we saw possums climbing the tree near the hut we were staying in, and the ingrained response to protect the kiwi kicked in, as somewhat of a birthright. Armed with the spears that our kids had whittled that day, we attempted to knock the possum from the tree. Our friends, Australian as it happens — from the land of the protected possum — were surprised to see us so 'bloodthirsty,' as they put it. We were driven by sheer shock that a predator was in such proximity to our beloved national bird.

And so I come to think of this bird and its status as a national symbol of the underdog who can overcome adversity and punch above its weight. It is not the bird I think of when I think of home. Instead I think of the tītī that Mihi-ki-te-Kapua sang of, I think of the kererū that my whānau recall eating, and more latterly I think of the myna bird — introduced yet uninvited — thriving in the paddocks surrounding Nan and Koro's. It has more of a presence in the whārua than the kiwi.

•

Inspired by these photos, and other things, I have signed up to an online weaving course. Well, I did so a couple of months ago and haven't woven anything yet. It's easier to watch videos than it is to get up and cut harakeke. There feels to be an invisible barrier between the watching and the doing, something stopping me. I recall the words of an Aboriginal friend who once said to me that trying to learn your language can be more painful than never having had it in the first place. This is what my barrier is, a familiarity with a knitting needle that I don't have with harakeke. It's a barrier that feels like a failure.

•

On a recent drive between Rotorua and Rūātoki, Dad reminded me of another connection that our people have to the Far North: te waka Mātaatua. Our whānau took a haerenga to the Far North at the invitation of my Koro Raniera, my Nanny Nan's husband. We visited Cape Reinga, the Bay of Islands, heard of the hāpuku that were so plentiful that boats would be weighed down by their day's catch. We stopped at a school, where the koroua hosting us pointed to the harbour and showed where Mātaatua lay. He quipped that if we knew the right karakia to uplift the waka, we could have it back, adding, 'If anyone is to know the right karakia, it would be Tūhoe.' We swelled with pride.

I do not know the karakia.

•

I do not know the karakia. I do not hold a mussel shell.
I do not wear a kākahu of kiwi feathers.
I have our kōrero. I have my whānau. I have our whakapapa. I have our songs.
These are my feathers.

A MANTLE FOR OUR SHOULDERS

Cassandra Barnett Raukawa, Pākehā

Fly, kiwi

The kiwi feathers bring a twinge of recognition. Short brown strands splaying loosely from thin shafts, spaced out like teeth on a wide-toothed comb. But soft, like the lower tufts of a peacock or ostrich feather. But soft-rough. Like hair. Even seen through a hung plastic bag. Even upon a death-slumped body. Even compressed against that casing, I know these feathers.

•

Eugene trimming the feet off a bird, using what appear to be loppers. The image shines with an unmistakable radiance. It's nowhere in particular. It just emanates from everywhere — the bird's vulnerable post-life repose, the sombre attention on Eugene's face, the additional gloved hands cradling beak or leg, the stillness. It's the opposite of violence.

Yet it's violence.

•

Three gloved hands attending to a bird's neck, one holding an unidentifiable instrument. A question.

•

Two feet. One severed head. So very small in proportion to its beak — not to mention its lost body. An answer. My eyes cry.

•

A kiwi, whole again. A kiwi still holding its cloak, still wearing its life feathers.

Suddenly it's too whole. Abundantly, almost wastefully, whole. I have accustomed myself. I see not just death, but something else in waiting. A vessel of wealth, abundance, generosity. Ready to give itself up, completely.

But my eyes betray me, crying again. The limp body still holds the pose of dying.

•

The wholeness carries me not just forward but back, to vulnerability, aliveness, a self-sovereign inner life — and to empathy or pity.

A danger-moment. Which way will the feelings carry me?

•

Final image, close-up: a kiwi head in repose. The beak arcs up to the top corner, calmly pointing somewhere. Though we can't see outside the frame, we feel the body intact.

•

Nothing is added. No weight of history or symbolism. If anything, lightness. Poise.
I want to say, the great lightness of surrender. Of transition. Of liberation.
I'm projecting.
But it's true too.

Pere wheua, patui wheua, punctum

I come to this project as a pīpīwharauroa.[1] This is not my rohe, I am not of this hapū. Yet in looking at these wounding photographs of kiwi being pelted at a Northland wānanga, I feel their grave relevance. I am of this country, I am Māori, I have named my son for the huia — a bird already lost to us. I have not even seen that many kiwi. Almost all of my live kiwi memories are framed by heavy doors, red-light switches, cool, hard corridors and glass partitions. The kiwi I've seen have been screened and separated from my body, my needs and circadian rhythms. We've coexisted, briefly, in cognitive and physical dissonance at zoos and wildlife centres. I don't know what it feels like to share an environment, let alone a habitat, with a kiwi. I don't know what it is like to inhabit the same auditory field, the same thermal envelope. And I'm not sure I've ever properly heard the sounds of a kiwi going about its business. My knowing of kiwi life is thin.

Yet the images prick me like a pere wheua, or a patui wheua, indeed sometimes made of kiwi bone. The images matter to me. They prick me as they stitch together so much that matters right now, and thus they stitch *me in,* too.[2] They stitch me in to the point that I think I may need to reconsider my cuckoo status, may need to decide that I also belong to this project. But I can't, not yet.

Perhaps these photographs are so poignant because they point to so many deaths. The deaths of the individual creatures — already too many, too disturbing in number (a hundred kiwi corpses in the Bay of Islands DOC freezer since 2013[3]). The hovering possibility of the death of the species;[4] certainly the death of the species in its unmanaged, 'wild' existence. With that, the death of a (Pākehā) national icon — so iconic that its extinction could symbolise the death of an era or phase in (Pākehā) national identity.[5] The death of our forest kaitiaki, of one more living practitioner and living embodiment of ecosystem kaitiakitanga[6] — and all that *that* death represents. The death of customary cultural practices associated with kiwi, from raranga to whakairo — more taonga that will not have survived the protections of colonisation and settler conservation practices.

The statistics (many recounted in Cinzia Vestena's text in this book but also freely available online) are more than unsettling; they induce despair. The sense of ring-fenced populations of kiwi dependent on vigilant predator-control (including car, cat and dog control) not just by DOC but by active, concerned local communities. That most chilling statistic: the death rate of newborn chicks in non-controlled areas is 94 per cent (due mainly to stoat and cat predation).[7] The utter dependency of kiwi, for species survival, on *our own voluntary vigilance*. When we all know how hard it is to sustain vigilance through the fluctuations of policy, resourcing, human population migrations, tides of public and private feeling . . . How often are you setting your traps?

Despite all this I think the source, the puna, of these photographs' poignancy lies elsewhere. The sharpest pricking comes from this: the people in them are quietly working against that tide. They occupy a space where, though the larger deaths may be coming, *they have not happened yet*. A space where we can still practise our ways.

So while it is tempting, feeling the pricking and stitching and mattering, to fly straight to the heartbreak of kiwi extinction and stop there at a now-familiar verge of despair, that is not what these images are asking for. And it is not what they project. Rather, they project deep resilience. Deep intimacy. And deep commitment to the ongoing economy and ecology that is kaitiakitanga.

Resilience, intimacy

Since the Kaikōura earthquake of 2016, also strongly felt in Wellington, my flatmate and I have been speaking about post-disaster survival requirements in our residential suburban setting. Our house is in Ngaio, a suburb in the hills at the northwestern edge of the city. Behind us, Kaukau maunga teems with gorse and rabbits. One of the things we've talked and talked — but only talked — about is the viability of living off those rabbits.

We're both woefully unskilled. My flatmate was overlooked for the 'training' because she is a girl, despite a West Coast upbringing in a bush-capable family. I was a soft city kid for most of my formative years, despite farm-raised parents and a hapū-ful of bush-capable whānau I wouldn't meet until I was in my thirties. Knowledge loss only takes a generation.

We figure we could trap a rabbit, and probably donk it on the head pretty easily too. But the key is skinning and gutting the thing, and here our conversation always ends. 'Yeah,' one of us will say, 'must get my brother to come give us a workshop.' And we'll turn back to stirring our pasta or our supermarket meat freshly divested of its plastic skin.

Kaitiaki, manaaki

The old ecosystem is broken. In life, kiwi are no longer plentiful enough to keep the insects down — and the insect community is changed anyway. In death, kiwi are no longer plentiful enough to become food in turn for the earth, plants, trees, insects and moths, let alone to be killed for their feathers. Yet their mana within that ecosystem is preserved — sort of — in the present-day Treaty-endorsed relationship between hapū-custodians and 'found' dead kiwi. Their stay in DOC freezers serves as a kind of portal between worlds, holding time for the conversations that will facilitate their passage. If this gathering in freezers is a visible marker of something deeply awry, once they are transported into the hands of their kaitiaki things change. Out of the freezer, into a different world — into participation in a heterogeneous economy. Into the active, caring, engaged, respectful, *alive* practice of kaitiakitanga we witness in these photographs.

The wānanga led by Tiwai Rihari-Rawiri and Raewyn Ormsby-Rihari, both connected with Ngāti Torehina ki Matakā, allow these kaitiaki of Northland brown kiwi to practise customary skills: preparing, pelting, storing, plucking and weaving kahu kiwi. They create an opportunity to teach these skills to future generations of Ngāti Torehina ki Matakā, keeping precious mātauranga alive. They honour the killed kiwi and uphold their mana, creating taonga from their feathers. And they invest Ngāti Torehina ki Matakā with taonga that will in turn uphold the mana of their people on occasions when a kahu kiwi is called for.

The whole cycle is there, sitting quietly behind the photographs. What tears fall, fall from aroha for the beauty of this system so imbued from beginning to beginning with care for every participant in the process. The seeming brutality is folded into the manaaki, held in an embrace that gives it its place. It is a quite different relationship from capitalist ownership; indeed, the feathers and kahu kiwi can be neither owned nor sold commercially.[8] Such cycles are part of the kaitiakitanga economy I see at work here and long to learn from, even at this distance. But it's no mean feat. To engage in an ecosystem as a *participant member* of that ecosystem — a kaitiaki — entails all of this: handling other living creatures with *this* level of full-bodied, full-hearted commitment. We may have to cut heads and feet off. We may have to cry. We may need karakia, and people with the right knowledge for this mahi. We may have to feel alive, before we die.

The images bear all this. The cost of kaitiakitanga is great. It is a lot to bear.

As the pūrākau tells us, back then, only the kiwi could bear it.[9] Could *I* bear it? Can humans — and I mean, now, not just the hapū looking after kiwi within their rohe, but all of us — sacrifice our flight, our treetop views, our sunlight, for the sake of maintaining the forest floor? So many problems in our country alone, including kauri dieback, tell us no. No, we can't. And I come back to the question I raised at the start, the question of whether *I* belong to this project. I know I belong to this problem. I know that somehow the difficulty of dealing with death is linked to the difficulty of sustaining life. And I somehow feel, though I'm distant and need to learn different intimacies with different rohe, that these images and the feeling inside them hold an answer I must listen closely to.

Pelting, skinning

It's just one small skill in a great string of essential, lost life skills. But in some ways it feels more essential even than the killing, which after all can happen by accident, can be done 'for you'. To skin a creature, you have to get your hands dirty. It requires intention, commitment, grit. A kind of violence.

It also confirms beyond doubt that the death will not be reversed. Life is not coming back, not as it was known by this being anyway. And it seems to mark the transition from one value system, one economy, to another.

We've learned our ignorance within mainstream urban Pākehā culture. It's part of larger biases, maybe — we would rather be separate, everyone's skins intact, preserving our distinction from the mess of the world over there. Even though the rabbits are pests and we've arrived, rationally, at eating them being a good idea, we just don't know how to make it happen. I should add that I've tried. I've mangled fish trying to extract hooks and kill them. I've bungled guttings and pluckings before. Without the tikanga, it's not pretty. How to turn ourselves into survivors, subsisters? Participants in an ecological community.[10] *We long to become these things. We have to overcome ourselves, in order to begin.*

Insides, insideness

The care with which these kiwi are divested of their skins feels so singular. It holds within it the beforelives and the afterlives. We see it on the faces of the weavers and their whanaunga. We see it in how they wield their tools. It says: We want to do this as much as we don't want to do this. We do this for all of us. We do this because we thank and honour you, kiwi. We do this because we want the world that also nurtures you to continue living. We want to all be a part of it.

Death is death. But having performed their mana-bestowing earthly role, the mana of the kiwi lives on in their feathered cloaks. The tender handling to which their bodies now surrender, even in dismembering, expresses that mana.

It expresses respect and gratitude — and mourning for their scarcity. It is the opposite of the many variants of the trophy photograph — though perhaps pride has some place when killing for conservation, as it does when killing for kai. When rendered through the weavers' skill into human cloaks, that kiwi mana lends itself to new wearers. And the new wearers, feeling it, remember to check on the kiwi. And onwards.

Thus, in death, the kiwi permit a customary practice to live on — so long as there's not *too much* death. So long as they are reproducing and surviving into adulthood. This kaitiakitanga practice requires kiwi to live. It's a reciprocal dependency, a mutual conservation, an ecological relationship that we're all inside of.

Fences, borders, trappings

The margin-dwelling humans like myself, between city and green belt, put out their traps. Small hubs of intensified human interest organise predator-free rallies, zones, actions. Import and distribute rat traps to homes in the 'hood. Try to stimulate vigilance through neighbourly bird counts, rodent catch counts, bait forums: 'What's worked for you this week? We had a win with Whittaker's almond chocolate!' 'Three big ones on peanut butter!' 'The bacon fat yielded nothing — perhaps they smell a rat? Lolol.'

Neighbours spy on neighbours. Who is the weakest link? The ageing Irish immigrant on her bad hip, struggling to walk the path to her car? The two solo mums flat-sharing with their toddlers — toddlers whose fingers almost want to feel the kapow-bite of the American spring-loaded mechanism? Some right on the bushline skite: 'We got seventeen this week, seventeen!' Others rejoin: 'We've moved on from that free trap — fiancée got me one of those auto-loading Good Nature ones for my birthday. Don't have to reset it for weeks! We're eradicating those bastards.'

Eradicating, entering

Are you? Are we? Even if it reaches critical mass in a community, can small-scale unilateral human vigilance simulate the effects of a balanced ecosystem? The border zone remains porous. A string of traps is no substitute for a predator fence. What if someone new moves in? Lines and separations have their place in a war machine. But in all this pest-warfare I smell a rat. *Then what?* Even if all the possums and rats and mustelids are eradicated by 2050 — it won't happen — then what?

Would every person then re-enter the balanced ecosystem as a participant-kaitiaki, and not just a sniper at the border? Or do the non-kaitiaki just keep trapping while hapū-kaitiaki keep processing the native casualties? Are Māori alone going to bear the full meaning and burden of this mahi kaitiaki? Will we need to bear it alone forever? Or will everyone else step up to transforming themselves, imaginatively committing to their part? What is the new 'balanced ecosystem'? A face-off between pest-eradication zeal and kaitiakitanga?[11]

I'm seeking an understanding of kaitiakitanga that applies to the ecosystem at large, but is very small and personal. Something with the commitment of trappers and hunters minus the worst shades of blood-thirst and trophy-ism. Something that holds onto respect and care for every creature. Something that acknowledges the cultural and spiritual cost of all this killing. Something that recognises the scale of the surrender we require, even from the creatures our European tūpuna brought here. Something that can feel the grief of that, somehow commit to them in death as in life, *and* to the wider ecology, and start working. What if we start needing *everything* we have, finding a place for it and learning what it has to teach us?

Impossible questions, maybe.

Rāhui worked, when done right.[12] It's too late for that now.

Keepers

> [Sharonda] imagined them still full of organs, blood, sinew, life. When she saw it she imagined it coming off. A knife in between fat and skin, and then the pull. It was the pull that did it. The weight the puller had to put behind their work. It didn't slide off, it resisted, like anything alive would pull back. 'No,' it seemed to say, 'no, I want this. I want to keep it.'[13]

The cow-skinning imagined by Sharonda in Pip Adam's novel *The New Animals* is alienated from the flows of reciprocal conservation or kaitiakitanga. It is a fast-fashion pelting at the brutal, globalised, noncircular end of a capitalist resource-exploitation economy. The skin passes swiftly from self-sovereignty to use-value to exchange-value. Now we will profit from you. Sharonda is right to care about this. And what if the skin puller also cared, and the leather was not for profit, and all were embedded in an economy, and practices, of kaitiakitanga? Would the cowskin still seem to say 'No'? The kiwi do not seem to say 'No'.

Within te ao Māori and kaitiakitanga, the concept of 'exploitation' is alien, a misfit kupu. Exploitation tends to mean using *up*, use-to-depletion, use-without-renewal, use within cradle-to-grave rather than cradle-to-cradle resource flows[14] — short-sighted uses that have facilitated so many extinctions.[15] In *Mountains to Sea: Solving New Zealand's Freshwater Crisis,* a number of contemporary Māori writers describe what kaupapa-Māori-based ecological economies would look like. Tina Ngata gives words to the hunch I've been feeling before these photographs: that intimacy is at the heart of kaitiakitanga; that you can't be a kaitiaki from a distance.[16] Meanwhile Paul Tapsell and Alison Dewes reiterate how an agenda of economic growth without limits, disregard for indigenous knowledge systems, and wilful blindness towards environmental degradation has brought Aotearoa to a crisis of 'pollution, poisoning and extinction'.[17]

But as I've already said, I don't want this text to focus on despair or grief. The beauty of these images, and of this wānanga, is that they are *not* despairing. They bring us close to grief, they permit mourning, but they are bathed in light. They bring the kiwi deaths out of cold storage, literally, and launch them on their next journey. Into the heavens as stars.

Cloaking

These images are reminders of how to care, how to love — of the burden, the difficulty, the commitment, the action of that. The action of rising up off our seats and doing the mahi, wholeheartedly. Putting our sadness and our aroha to work. Not just because something beautiful will be produced. Not just so the death of these kiwi is less wasteful, less in vain. Not just to preserve our culture, pass the mātauranga, the ethos and the practices of kaitiakitanga — and the mana of the kiwi too[18] — down to our rangatahi . . .

But because in this way, through singular, here-and-now action practice, we reel the far-flung strands of despair and hopelessness back in. We remember we are inside, not outside, the ecosystem, sick as it is.[19] We activate our insideness, which is to activate our small agency. We come back into our living, connected bodies and we *do* and *live* our connections — our ancient connections. This doesn't just activate hope and activism for a better world. This puts us back into a world — however small — that is *well.*

Despite the humility, grace and gentleness of the images — but actually because of it, too — I find myself arriving at a manifesto of sorts. As kaitiaki we must find and honour our own charges, our own wards. Māori and non-Māori alike must take our places as *ecological participants.* We must start getting our hands dirty (and I don't just mean trapping), stop living separated from our own relations to things. Must develop our guts for this. *I* must dig deep into *te tau o taku ate* — my liver, the seat of my feelings — and awaken the connective, activating grit and aroha for the job.

I'm working on this not being too much. I'm working on feeling this not as fantasy or romanticism but as a survival necessity. *A reality.* I'm still looking for someone to teach me to pelt the rabbits and possums my European tūpuna brought here, and someone else to tell me what I can and can't know about our neighbourhood manu and what they need from me (besides the trappings), and someone else to tell me how to honour the rats in turn.

I'm working on bringing the ecosystem inside of me and myself inside of it.

I'm thinking about kiwi. Not Kiwi, symbolic, emotive, generic, abstract, emptied, flat. But kiwi, *these* kiwi, kiwi that once lived and are now dead. Kiwi on the threshold: this side of the gateway, the *pare,* then that. Warm, quivering. Cold, dead. Looking at them I cry.

Everything is about these thresholds. First living, now dead, now living on, but also, definitely, dead. I'm looking at the tender, still kiwi who have given up all they had and preserved so much in the process. I'm thinking again and again about surrendering to a world that is well, and how hard but how urgent it is.

It may all be too late.

I'm doing it anyway.

---

1 The pīpīwharauroa places its eggs in the nest of the riroriro, where its babies are incubated and hatched by these other mothers.

2 French theorist Roland Barthes calls such pricking the 'punctum' of a photograph. We don't need to know this to feel it.

3 Media release, Department of Conservation, Kerikeri, 2017.

4 As Cinzia Vestena discusses on page 39, the IUCN Red List gives Northland brown kiwi a Vulnerable rating and they face a high risk of extinction in the wild.

5 This last death would be felt more painfully by some than others. Many Māori have already suffered deaths of identity in these lands.

6 See kaumātua Hugh Rihari's retelling of the kiwi pūrākau, 'Te tūtakitanga a Kāhu rāua kō Kiwi', on page 74.

7 See page 40.

8 Māori Marsden: 'Ownership of property in the pre-contact period was a foreign concept [...] the resources of the earth did not belong to man but rather, man belonged to the earth,' in *The Woven Universe: Selected Writings of Rev. Maori Marsden,* Te A. C. Royal, ed. (Otaki: Estate of Rev. Maori Marsden, 2013), 67.

9 See page 74.

10 I may be Māori while my flatmate is Pākehā, but in this we have more in common than not. I need to be truthful on this point: I lack the skills as much as she does, and she is as eager to learn as I am.

11 This is a small fragment of a larger national debate requiring policy and legislation changes; my focus for now is on what personal shifts we are capable of through the mere process of witnessing images like these.

12 Māori Marsden describes rāhui as 'a prohibition or ban instituted to protect resources'; an area

would be 'placed under the tapu of rāhui in order to allow it to regenerate' (Royal, ed., *The Woven Universe,* 69), and giving 'tangata whenua [...] the right to control the access of other people and their own tribal members to the resource' (71).

13 Pip Adam, *The New Animals* (Wellington: Victoria University Press, 2017), 11. Courtesy of Victoria University Press.

14 'The institution of rāhui was designed to prohibit exploitation, depletion, and degeneration of a resource and the pollution of the environment to the point where the pro-life processes latent within the biological and ecosystems of Papatuanuku might collapse.' (Marsden, in Royal [Ed.], *The Woven Universe,* 70).

15 A new wave of alternative, circular economies are being touted now as the new global solution to waste, environmental degradation and sustainability. Without going too far into that, I want to differentiate between any global, happily entrepreneurial and purportedly 'new' solutions to these problems and wholistic economies such as kaitiakitanga that have long been embedded in Māori (and other indigenous) practices.

16 T. Ngata, 'Wai Māori', in *Mountains to Sea: Solving New Zealand's Freshwater Crisis,* ed. Mike Joy (Wellington: Bridget Williams Books, 2018), 10.

17 P. Tapsell and A. Dewes, 'One World, One Health, One Humanity — Whenua, Rongoā, Tangata', in *Mountains to Sea,* 31.

18 Though I do think about my own graduation this year at Te Herenga Waka marae, and how I walked uncloaked — though still wrapped up in aroha! — because that is how my history has woven my life. I think: the girls of this hapū will not need to walk uncloaked! And I'm glad for them.

19 If there are to be lines and predator fences, we will need to figure out where we sit in relation to them. Are Māori inside and Pākehā outside — as the predators' predators? Is this a dichotomy we want to see enduring?

TE TŪTAKITANGA A KĀHU RĀUA KŌ KIWI

Hugh Rihari Ngāti Torehina ki Matakā, Ngāti Mau

I ngā wā o mua, i muri i te taenga mai o Kāhu ki Aotearoa,
ka tūtaki a Kāhu ki a Kiwi.
Ka kite a Kāhu i te iti o te tinana o Kiwi,
kua ngaro ngā pākau, kua roa te ngutu timotimo me te nunui o ngā matimati hao.
'E Kiwi, e pūhaehae ana koe ki a au? Te nui hoki o tōku tinana,
te nui o ōku pākau me ōku huruhuru? He manu kaha ahau ki te rere,
he manu kaha ahau ki te patu kararehe hei kai mā āku tamariki.'
Ka whakaaro a Kiwi ki ngā kupu a Kāhu mō tētahi wā.
'Āe, he iti noa o mātou tinana, kua ngaro kē o mātou pākau,
kua tupu roa o mātou ngutu timotimo, me o mātou matimati hao.
I te taenga mai o ōku tūpuna ki Aotearoa, he manu rererangi mātou.
I muri mai i te taenga o ōku tūpuna ki Aotearoa ka karangahia e Tāne-mahuta
he wānanga mō ngā manu katoa.
'E ngā manu katoa, kua pā mai tētahi mate kino ki āku tamariki,
ngā rākau o te ngāhere.
Ko ngā kākano o ngā rākau, āku mokopuna,
e kainga ana e ngā ngārara o te whenua. Me heke tētahi o koutou
ki te noho tūturu ki te whenua hei patu,
hei kai i ngā ngārara e kai nei i āku mokopuna.
Ko te manu e whakaae ana ki te heke tūturu ki te whenua,
e kore rawa rātou e hoki ki ngā mangamanga o te ngāhere.
Ā te wā ka ngāro o rātou pākau, e kore anō e rere i te rangi.
E Kāhu, kō ōu tūpuna me te nuinga atu o ngā manu
i whakakāhore i te karanga a Tāne-mahuta, ko ō mātou tūpuna anake i whakaae.
Ko ō mātou huruhuru he huruhuru tino rangatira.
Ka mate ana mātou, ko ō mātou huruhuru hei korowai mō ngā tino rangatira o te ao.
E Kāhu, kua tino iti koutou.
Ā te wā koutou ka ngāro, engari anō mātou ngā kiwi,
kei konei tonu mō ngā whakaheke kei te haere mai.
Ahakoa ngā piki me ngā heke kei konei anō mātou ināianei,
me te wāheke kei te haere mai.'

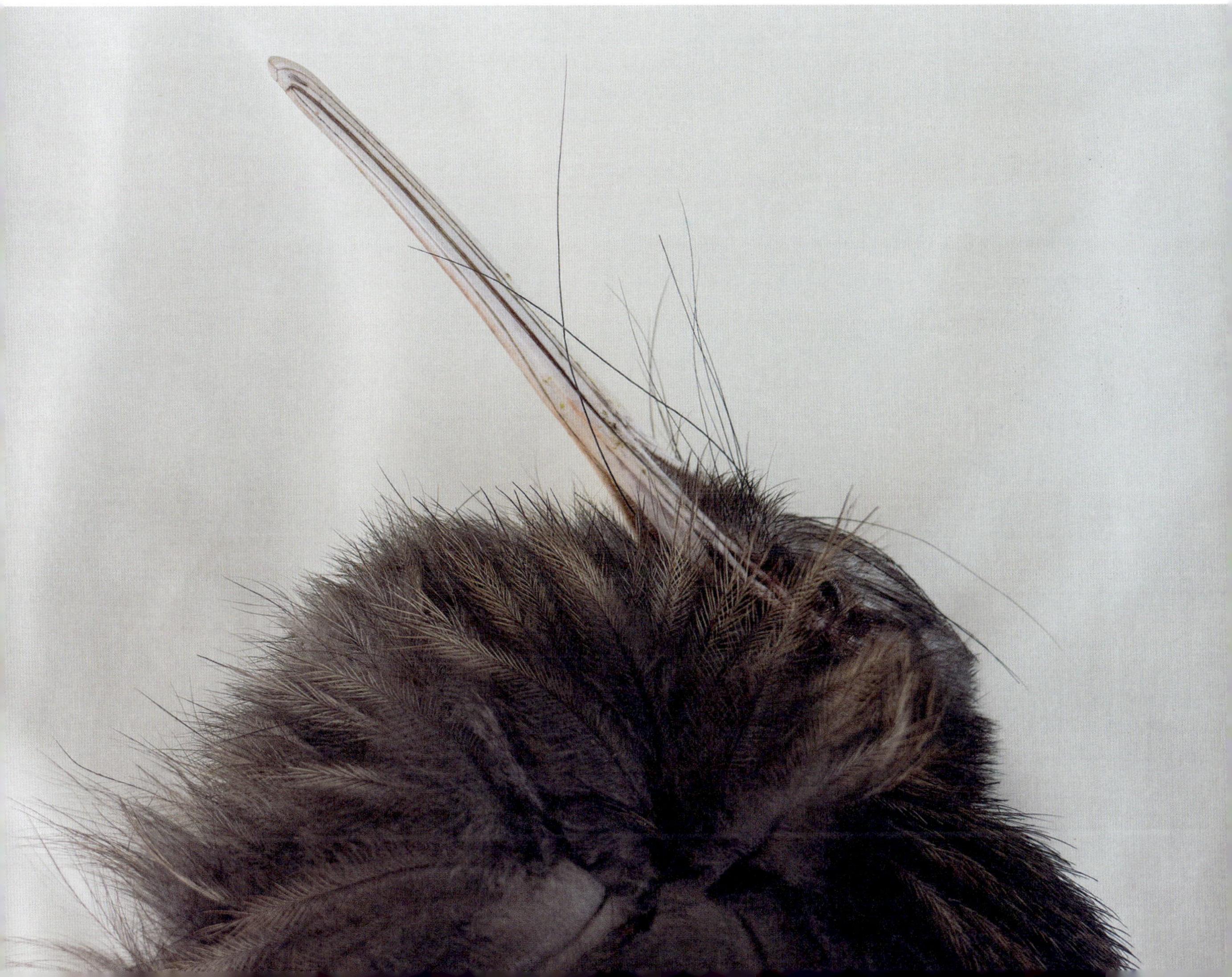

Pictured

Dolly Rihari, Ngāi Tūhoe — pages 24, 45, 46 and 54

Raewyn Ormsby-Rihari, Maniapoto, Ngāti Torehina ki Matakā by marriage — pages 24, 25, 26, 45, 46, 52 and 54

Tiwai Rihari-Rawiri,Ngāti Torehina ki Matakā, Ngāti Pare, Ngāti Mau — pages 45, 46, 52 and 54

Te Aroha Matenga, Ngāti Torehina ki Matakā, Ngāti Pare, Ngāti Mau — pages 27, 46, 52 and 54

Karyn Tuapawa, Ngāti Porou — pages 24, 46, 52 and 54

Aroha Rihari-Reihana, Ngāti Torehina ki Matakā — pages 46, 52 and 54

Tyler-Jo Rihari, Ngāti Torehina ki Matakā — pages 24, 46 and 52

Roxxanne Rihari, Ngāti Torehina ki Matakā — pages 46, 52 and 54

Eugene Hansen, Maniapoto — pages 26 and 27

Felix and Asher Hansen, Maniapoto — page 54

Contributors

Cassandra Barnett (Raukawa, Pākehā) lives in Ngaio, Te Whanganui-a-Tara. She is a solo mum and a writer, with fiction, essays and poetry published in *Landfall, Turbine Kapohau, ATE: Journal of Māori Art, Counterfutures, Brief, OraNui, Black Marks on the White Page* and in many arts publications, and broadcast on Radio New Zealand. She has completed a PhD on contemporary art in Aotearoa (University of Auckland) and an MA in Creative Writing (Victoria University of Wellington). In 2019 she was writer in residence at Zealandia, guest editor of *4th Floor journal* and a finalist in the Pikihuia Awards for Māori Writers.

Rolien Elliot has worked for Te Papa Atawhai Department of Conservation since its inception in 1987, as a ranger and manager in various locations across New Zealand before taking on the role of Bay of Islands Area Manager in 2007. She is Strategic Projects Manager for Tai Tokerau, where she is involved in regional conservation projects. Rolien presented evidence on behalf of the Crown at Waitangi Tribunal claim hearings Wai 262 Flora and Fauna and more recently the Wai 1040 Te Paparahi o Te Raki inquiry.

Jenny Gillam trained as a photographer in the early 1990s and gained an MFA (RMIT, Melbourne) in 1999. She is a multi-media artist interested in the socio-politics of ecology; aspects of current debate around ecological tensions; humankind's relationship with animals, and our place in the natural world. Her art projects are often produced collaboratively with other artists or with experts from other fields. Jenny is a Senior Lecturer at Massey University's Whiti o Rehua School of Art in Wellington.

Eugene Hansen (Maniapoto) is Raewyn Ormsby-Rihari's son. He is a Senior Lecturer at Massey University's Whiti o Rehua School of Art in Wellington. Focusing on co-authoring and working collaboratively, he has a long-term multimedia art practice exhibiting nationally and internationally. Eugene attributes his interest in collaboration to growing up in the remote rural Māori community featured in this book, where cultural production was modelled as an inherently collaborative embracing of mātauranga Māori.

Raewyn Ormsby-Rihari (Maniapoto, Ngāti Torehina ki Matakā by marriage) has been weaving for many years, initially with a loom then learning traditional Māori raranga. In 2010 she was a wānanga huruhuru manu participant, and she became an assistant tutor the following year. She is a registered kiwi advocate and permit holder. Raewyn was iwi monitor for the construction of the Rangihoua Heritage Park Walkway, Marsden Cross, Kerikeri, and has been involved in archaeological surveys at Heritage Bypass, Kerikeri, in 2007, Patunui Bay in 2012 and at Marsden Cross in 2013. Raewyn has a contemporary kākahu on permanent display at Rewa's Village in Kerikeri.

Hugh Rihari (Ngāti Torehina ki Matakā, Ngāti Mau) is a descendant of Wharemokaikai who, with his older brother, Te Uri o Kanae and others, welcomed the missionary Samuel Marsden and his followers to Hohi in December 1814. As a kaumātua he has worked closely with many communities in the Bay of Islands for a number of years. He leads, liaises and advises on resources, development and kaupapa, and advocates for tangata whenua, working closely with Far North District and Northland Regional councils, central government, Te Papa Atawhai Department of Conservation, Heritage New Zealand, and others.

Tiwai Rihari-Rawiri (Ngāti Torehina ki Matakā, Ngāti Pare, Ngāti Mau) has been a raranga tutor since 2010 and regularly holds wānanga at Aroha Island, Bay of Islands. She gained a diploma, Te Pōkairua Ngāpuhi-Nui-Tonu Te Raranga in Ngā Mahi a Te Whare Pora (Māori Weaving), from NorthTec in 2013. She has been involved with Te Papa Atawhai in the wānanga huruhuru manu since 2011 and a registered kiwi advocate and permit holder since 2016.

Toi Te Rito Maihi (Ngāi Te Ipu, Ngāi Te Apatu o Ngāti Kahungunu, Ngāpuhi) has been a full-time artist since 1976, exhibiting widely. She is a painter, printmaker, author and fine weaver of tāniko, kākahu, whāriki and kete whakairo. Her work is held in the collections of the Museum of New Zealand Te Papa Tongarewa, Auckland Art Gallery Toi o Tāmaki, MTG Hawke's Bay Tai Ahuriri, and Waikato Museum Te Whare Taonga o Waikato. She was a major contributor to *Whatu Kākahu | Māori Cloaks,* a publication that accompanied the *Kahu Ora* exhibition held at the Museum of New Zealand Te Papa Tongarewa in 2013. A second edition of the book was published in 2019.

Cinzia Vestena is Biodiversity Assets Ranger at Te Papa Atawhai Pēwhairangi Department of Conservation Bay of Islands. She works with community, tangata whenua and other government agencies in developing a management plan for kiwi populations and provides advice to community groups on threatened-species management and translocation. Cinzia moved to Aotearoa from Italy in 2000 and is now based at Matauri Bay, Whangaroa. She gained a Postgraduate Diploma in Environment from the University of Auckland in 2002.

Matariki Williams (Ngāi Tūhoe, Ngāti Hauiti, Taranaki, Ngāti Whakaue) is a Curator Mātauranga Māori at Museum of New Zealand Te Papa Tongarewa. With Bridget Reweti she co-founded and co-edited *ATE: Journal of Māori Art.* Her writing has appeared in print and online publications including *frieze, The Pantograph Punch* and *The Spinoff.* With Stephanie Gibson and Puawai Cairns, she is co-author of *Protest Tautohetohe: Objects of Resistance, Persistence and Defiance,* published in 2019. Matariki is also a trustee of the arts writing website Contemporary HUM.

Translations

page 5

Sitting and listening to kiwi calling at night, this is what you hear. The still of night is pierced by the shriek of a male kiwi hoping to attract a partner.

'Is there anyone out there?' calls the male kiwi.

The night falls silent again, then in the distance, the silence is broken with the piercing reply of a female kiwi, 'I am looking for a partner, could we meet?'

'Yes,' replies the male kiwi, 'we must meet. I'll keep calling to give you my location.'

'Yes,' she replies, 'I will meet you soon.'

page 74

WHEN KĀHU MET KIWI

Hugh Rihari Ngāti Torehina ki Matakā, Ngāti Mau

In our distant past, after migrating to Aotearoa, Kāhu met Kiwi for the first time and said, 'I am the biggest and most powerful bird; my feathers are strong and can carry me all day without rest. My size allows me to kill large animals to feed my family. You, Kiwi, are small in size, and cannot fly. Don't you wish you were big like me and could fly?'

Kiwi took some time to contemplate a reply. 'Generations ago, my ancestors could fly like other birds. It was at this time Tāne-mahuta summoned all birds and explained that the forest was dying because the berries that fall to the ground were being eaten by insects living on the forest floor so the forest was not being replenished. Tāne-mahuta asked who among the many variety of birds present would volunteer to live permanently on the ground to protect the fallen berries and new seedlings.

'It was only my ancestors who volunteered to live on the forest floor to kill and eat the insects which were responsible for the death of the forest.

'My ancestors understood that once on the ground, they would never fly again, and over time our bodies would change and our once short beaks would become long so as to reach the insects deeper in the ground, the claws on our feet would adapt and become longer and stronger, and our wings, without use, would wither away.

'We may not be big, strong and able to fly high over long distances like you can, Kāhu, but in our kingdom we feel big and strong. What we do, and how we live, we are not only saving our forests, but also the habitat of all the birds that nest there and eat the many varieties of berries from the trees, who would otherwise have died had we not been keeping the forest alive.

'While we live we protect Tāne-mahuta's forest, the cloak of his mother Papatūānuku, and when we die his children, the people of the land, honour our service and make cloaks from our feathers for their leaders.

'As kiwi, in the future we will endure all forms of hardships and population fluctuations, but we will survive, and be protected; your descendants will all but disappear.

'Kāhu, there are few of you, and in time there will be less of you. We may not have the attributes you have, but we will continue to be around for generations to come.

'We are still here today, and will be here into the future.'

Glossary

aho weft, cross-threads of weaving

ātaahua grace and beauty

autui cloak pin/s

haerenga journey

hapū subtribe, kinship group

hāpuku *Polyprion oxygeneios,* groper

harakeke native flax plant

harirū handshake

hīkoi journey

hui gathering, meeting

huia *Heteralocha acutirostris,* one of the New Zealand wattlebird species, now extinct

hukahuka Thrum, tassel of two strands, used in the ornamentation of korowai

huruhuru pelt / feather/s

Io supreme being

iwi tribe, extended kinship group, associated with a distinct territory

kahu kiwi kiwi feather cloak

kai food

kaiako tutor

kaikaranga woman who makes the ceremonial call to visitors onto a marae at the start of a pōwhiri

kaimoana seafood

kāinga habitat

kaitiaki person or group with the responsibility for taking care of an object or place, or the safe-keeping of expert knowledge

kaitiakitanga guardianship

kaiwhatu weaver/s

kākā *Nestor meridionalis,* a large parrot with a distinctive red-orange underwing

kākahu clothing, garment or costume

kākahu pokowhiwhi flax shoulder cape

kānga sweetcorn

karakia incantation, chant or prayer; frequently used as part of tapu (sacred or restricted) practices of making

kaumātua elder, a person of status within a whānau

kaupapa project, purpose / main body of a cloak

kete basket

kereru *Hemiphaga novaeseelandiae,* wood pigeon

kiekie *Freycinetia banksii,* a thick native vine that has long leaves with fine teeth crowded at the end of branches

kōauau pongāihu nose flute

kōrari native flax plant

kōrero conversation, speech, narrative

koro / koroua grandfather, male elder

korowai term sometimes used as a general name for a cloak; it is also a specific type of cloak ornamented with black twisted tags, thrums or feathers

kuia grandmother, female elder

kūkupa *Hemiphaga novaeseelandiae,* native bush pigeon (also known as the kererū)

kupu word, term

kuratini polytechnic

mahi work, practice

mana prestige, status, power

manaaki to support, protect, show respect

manaakitanga kindness, care

manu bird/s

manuhiri visitor, visiting group

marae complex of buildings and grounds; a meeting place for a particular iwi, hapū or whānau

maro traditional boys' kilt

Mātaatua one of the great Māori migration waka

matau fish hook/s

mātauranga knowledge, wisdom, understanding and skill; also relates to the passing on of traditional knowledge

maunga mountain, mount, peak

mau taringa earrings

mō āke tonu atu forever, eternal

muka the processed fibre of kōrari or New Zealand flax; muka is used in the creation of fine clothing and accessories

ngira needle/s

nguru a short flute made of wood, bone or stone, and played with the mouth or nose

noho marae a stay on a marae

Papatūānuku Earth, Earth mother and wife of Rangi-nui

pare carved slab over the door of a house, threshold

pareu traditional girls' kilt

patui wheua needle/s made of bone

paukena pumpkin

pere manu bird dart/s

pere wheua arrow/s made of bone

Pēwhairangi Bay of Islands

pīngao *Desmoschoenus spiralis,* a native plant with golden-orange narrow leaves that grows on sand dunes

pīpīwharauroa *Chrysococcyx lucidus,* the shining cuckoo, a migratory bird that lays its eggs singly in nests of grey warblers

piri loyal

poi light ball on a string swung in various movements to accompany singing

pōtae hat

pouwhenua long wooden weapon

pūkeko *Porphyrio porphyrio melanotus,* purple swamphen

pūrākau myth, story

pūtea money

putiputi flower/s

rāhui a temporary restriction on a site or resource

rangatahi younger generation, youth

raranga to weave / the practice of weaving

riroriro *Gerygone igata,* grey warbler

rito new flax shoot

rīwai potato

rohe district / region

rongoā medicinal plant/s

tamariki children

tāne man, men

Tāne-mahuta atua, or god of the forests and birds

tangata (pl) people, persons

tangata whenua indigenous people of Aotearoa, people born of the whenua, the land where their ancestors have lived and where their placenta are buried

tāniko the technique of knotwork or weaving using the fingers; often used on the border of cloaks

taonga a gift from the ancestors, an aspect of culture surrounded by knowledge, treasured item, valued person or thing; taonga can be tangible (such as huruhuru manu) or intangible (such as knowledge)

tapu sacred, restricted

tātua woven belt/s

tauiwi non-Māori person

te ao Māori the Māori world view, relating to language, principles and customs

Te Pō darkness, night, place of departed spirits

te reo the language, often used to refer to te reo Māori, the Māori language

te tau o taku ate the core of my heart

tikanga cultural practices and ways of behaving

timo beak

tino rangatiratanga self-determination, sovereignty

tītī *Puffinus griseus,* muttonbird

tohu mātauranga degree qualification / academic qualification

tohunga cultural expert. Often used in conjunction with an artistic practice to indicate expertise and knowledge in a particular area

tohunga whakairo expert carver

toi art, making. In contemporary usage 'toi' has come to represent all practices related to the visual arts

tokotoko walking stick, sometimes carved with patterns

tukutuku the woven panels placed between the carvings on the interior walls of a meeting house, originally made using natural materials such as kiekie, pīngao and kōrari fibre woven between them to form patterns

tūpuna (pl) ancestors, grandparents

wahine woman

wāhine (pl) women

waiata song, chant

waka canoe

wānanga to meet, discuss and share cultural and traditional knowledge

whaea mother, aunt. Often used when referring to a woman with affection and respect

whakaaro thought, way of thinking

whakaeke welcome onto a marae

whakairo to carve

whakapaipai ornament/s

whakapapa genealogy, lineage

whānau family, extended family

whanaunga relative/s, extended family member/s

whare house

wharekai dining hall

wharenui meeting house, main building of a marae

whāriki a type of woven mat used as a floor covering

whārua valley

whatu hand knotting, applied to the weaving of muka or flax fibre to make fine cloaks

whenu warp, lengthwise threads of flax weaving

whenua land / placenta

wheua bone

Acknowledgements

A deep sense of love and respect runs through both this book and the wānanga it documents. It has been a privilege to witness the generosity of the wāhine of Ngāti Torehina ki Matakā, your kaitiakitanga of one of our precious and vulnerable taonga, and your commitment to this tikanga that might otherwise disappear.
Hugh and Raewyn, your generous support, guidance and enthusiasm as this book has developed has been invaluable. I'm humbled by your trust in me.
Thank you to Raewyn and Tiwai, Hugh, Toi, Matariki, Cass, Rolien and Cinzia for your informative and heartfelt texts. They provide context and an understanding that can only be gained through lived experience. Thank you for bringing such rich and personal perspectives to these photographs.
Lots of love and respect to Eugene, for everything. And for your intellect and word-smithing as co-editor.
Thanks and much respect to Peter Miles, Susan Brierley, Hona Black, Adrian McCleland, Nicola Legat and Emily Goldthorpe for your invaluable expertise in producing this book.
And thank you to Anna Brown, Hiraani Himona and Huhana Smith for your initial support and encouragement.
We are incredibly grateful to Creative New Zealand. This book would not have been possible without support from the Suffrage 125 Fund.
Ngā manu ātaahua piri a Tāne, to mana mō āke tonu atu.

MASSEY UNIVERSITY PRESS

First published in 2020 by Massey University Press
Private Bag 102904, North Shore Mail Centre
Auckland 0745, New Zealand
www.masseypress.ac.nz

Design by Adrian McCleland, Waikato-Tainui

A catalogue record for this book is available from the National Library of New Zealand

Printed and bound in China by Everbest Printing Investment Ltd

ISBN: 978-0-9951230-6-9

creative nz
ARTS COUNCIL OF NEW ZEALAND TOI AOTEAROA